NOMOGRAMS
FOR
DESIGN AND OPERATION
OF
CEMENT PLANTS

NOMOGRAMS
FOR
DESIGN AND OPERATION
OF
CEMENT PLANTS

By

S. P. Deolalkar

Author of
Handbook for Designing Cement Plants

BSP BS Publications

CRC Press
Taylor & Francis Group

A BALKEMA BOOK

Deolalkar Consultants
B-32, Shanti Shikhara Apts.,
Raj Bhavan Road, Somajiguda,
Hyderabad 500082
India
E –mail : spdeolalkar@gmail.com

First published in India in 2011by,

 **BS Publications**
An unit of **BSP Books Pvt., Ltd.**

4-4-309, Giriraj Lane, Sultan Bazar,
Hyderabad - 500 095 - A.P.
Phone : 040 - 23445605, 23445688
e-mail : info@bspbooks.net
website: www.bspublications.net

Printed by
Sanat Printers
Kundli, Haryana, India

ISBN: 978-0-415-66577-3

Distributed in India, Pakistan, Nepal, Myanmar, Bhutan, Bangladesh and Sri Lanka by **BS Publications**

Distributed in the rest of the world by
CRC Press /Balkema, of Taylor & Francis Group
Schipholweg 107C
2300 AK Leiden, The Netherlands

NOMOGRAMS
FOR
DESIGN AND OPERATION
OF
CEMENT PLANTS

By

S. P. Deolalkar

Author of
Handbook for Designing Cement Plants

BSP BS Publications

CRC Press
Taylor & Francis Group

A BALKEMA BOOK

Deolalkar Consultants
B-32, Shanti Shikhara Apts.,
Raj Bhavan Road, Somajiguda,
Hyderabad 500082
India
E –mail : spdeolalkar@gmail.com

First published in India in 2011by,

 **BS Publications**
An unit of **BSP Books Pvt., Ltd.**

4-4-309, Giriraj Lane, Sultan Bazar,
Hyderabad - 500 095 - A.P.
Phone : 040 - 23445605, 23445688
e-mail : info@bspbooks.net
website: www.bspublications.net

Printed by
Sanat Printers
Kundli, Haryana, India

ISBN: 978-0-415-66577-3

Distributed in India, Pakistan, Nepal, Myanmar, Bhutan, Bangladesh and Sri Lanka by **BS Publications**

Distributed in the rest of the world by

CRC Press /Balkema, of Taylor & Francis Group
Schipholweg 107C
2300 AK Leiden, The Netherlands

This book also is for

Mrs Urmila

My Life's Partner

and

Proprietor of Deolalkar Consultants

Author

PREFACE

This Book, 'Nomograms for Design and Operation of Cement Plants' was conceived as a **complement** to my Book 'Handbook for Designing Cement Plants'. In that book, in section 8, the frequently required references and data were presented in CD format. A great many References were in 'active' Excel format to help the reader / designer to make fresh calculations to suit his own needs.

I had always found nomograms as quick and useful tools for quick estimation and calculations, just like a slide rule – which itself is a nomogram. Nomograms may not have the accuracy of calculators or computers but they are good enough for most purposes.

Charts and graphs are other ways of presentation and of arriving at results when there are more than one variable. They are also useful in their own ways.

Nomogram is an ingenuous way of presenting data (inputs) and of finding out required result in one or more steps. A mathematical equation is converted into curve/s and curve/s are converted into nomogram/s.

In principle, it is possible to construct a nomogram for almost all calculations required to be made in the design of cement plants. Choice would depend on the frequency of their requirements. For example, a designer has to arrive at capacities of main machines, their feeders and so on regularly. He also needs to find out space requirements of stock piles of limestone, clinker, coal etc., for given capacities in planning layouts. While such data can be tabulated, it can be presented more conveniently as nomograms.

In this book I have constructed and presented nomograms in Autocad for such applications so as to be useful for designers and operators of cement plants.

In presenting them I have roughly followed the format of my book –Handbook for Designing Cement Plants

The book has been divided in four sections.

Section 1 deals with 'Basics' which corresponds to chapters dealing with Basics in the book.

Section 2 deals with aspects related to converting volumes at n.t.p. into actual volumes and with calculating volumes of silos and stock piles commonly used in storing raw meal, limestone and clinker.

It also contains nomograms useful in calculating clear volumes of kilns, percent loading, power arm for calculation of power drawn etc.

Section 3 deals with process calculations

Section 4 deals with machinery.

Along with each nomogram is a text explaining the logic behind its construction and the basic assumptions. The nomogram as also the text furnishes details of scales used in their construction and also explains how to use it with an example. Thus I have tried to make them user friendly.

A large number of nomograms have 'log scales'. According to the situation, 5, 7, 10 and 20 cm represent 1-10,10-100. 100-10000 etc. Appendix 1 would facilitate reading intermediate values for different scales.

The accuracy of a nomogram depends on the scale used and hence there are practical limits to it. I may however assure the readers that it has the same kind of accuracy that a slide rule has.

I had to choose from a great many possibilities of constructing nomograms from practical considerations. After a great deal thought I decided to restrict the number to about 100 and hence had to delete a good many nomograms. This was of course a subjective decision and readers may have good reasons to be critical about some omissions. By way of further refinement, some nomograms could be combined into one.

As far as nomograms in the Fourth Section are concerned, Machinery Manufacturers themselves design nomograms for sizing and selecting their machinery. I have avoided using such ready made nomograms.

Psychrometric charts and charts used for calculating pressure drops in ducts due to flow of gases are other good examples of 'universally used nomograms'. These were included in the Handbook. Nomogram for pressure drop has been included in this book also.

I believe that the two together- the Reference Section of the 'Handbook for Designing Cement Plants' and this Book on nomograms together would prove to be of great practical use to all those engaged in the business of setting up of Cement Plants and in the business of operating them.

The Book is being brought out in the print form. A CD containing nomograms only is attached to the Book for the convenience of the readers. Nomgrams are in Autocad format which is readily available these days. However for the convenience of readers an autocad reader goes with the CD. There would thus be no difficulty seeing nomograms in autocad format and in using them actively. The procedure for drawing lines from new inputs to read corresponding outputs has been included in Appendix 2.

Feedback from readers would help me to improve on this book. Hence I look forward to receiving it.

- Author

ACKNOWLEDGMENTS

This Book is largely an outcome of my own efforts. Almost all nomograms have been constructed by me from first principles. I also evolved the excel programmes and the graphs that were the necessary first steps in the construction of nomograms.

I took the help of Mr. V. Ramkumar to put the nomograms on the Autocad. He had earlier helped me with the drawings for the 'Handbook for Designing of Cement Plants'. He has once again done the job diligently and meticulously. I am extremely thankful to him for the same.

In the construction of nomograms I had to use data published in various books., periodicals and manuals. I have mentioned the pertinent sources at the bottom of the text attached to the nomogram.

While I have taken every care to do so, should there be a few unintentional omissions, I seek forgiveness for the same.

In the early stages I took the help of Mrs. Bahar Tepan, Director, Softideas Pvt Ltd, to assess the feasibility of bringing out the book in CD format. The feasibility was established but from other considerations, the book is coming out in print form with a CD containing nomograms only.

The photo on the front cover is from the Brochure on 'Dry Process Kiln Systems' brought out by FLSmidth.

B.S.Publications, a unit of BSP Books Pvt. Ltd., readily agreed to bring out this Book also as they did the Handbook. They have done it with the same meticulous care and promptmess. As a result the Book has the same excellence in Quality as the Handbook. I am thankful to Mr. Anil Shah, Director, B.S.Books for the trust reposed in me this time also. Mr. Naresh Davergave, Production-in-charge of B.S.Publicatons and his staff have done an excellent job and I appreciate that very much.

- Author

CONTENTS

ABBREVIATIONS

Commonly used abbreviations are listed below.

1   **words**

dia.	diameter
rad	radius
st.	straight
thru	through
thruput	throughput
hr	hour
min.	minute
sec.	Second
pr.	pressure
esp	electrostatic precipitator
SCA	specific collection area
Nomo	nomogram
w.r.t.	with respect to
HGI	Hardgrove Index
BWI	Bonds Work Index
Sp.	specific
fk pump	fuller kinyon pump
temp.	temperature
ntp/NTP	at sea level and 0 ^{o}C

2   **Dimensions**

mm	millimeter
cm	centimeter
m	metre
km	kilometer

3   **Area**

mm^2	square millimeter
cm^2	square centimeter
m^2	square meter

4   Volume

cm^3 / c.c.	cubic centimeter
m^3 / cu.m.	cubic meter
nm^3	normal cubic meter (at sea level and 0 ^{o}C)
c.c.	cubic centimeter

5   Temperature

^{o}C	degrees Centigrade
^{o}F	degrees Fahrenheit
K	Kelvin absolute temperature

6   Weight

gm	gram
kg	kilogram
t/T	tonne (metric)

7   Velocity / speed

cm/sec	centimeters per second
m/sec	metres per second
r.p.m.	revolutions per minute

8   Density

kg/m^3	kilogram per cubic meter
t/m^3	tons per cubic meter
gm/c.c.	grams per cubic centimeter

9   Pressure

kg/mm^2	kilogram per square millimeter
t/m^2	tons per square meter
mmwg	millimeter water gauge

10   Specific Volume

m^3/t	cubic meter per ton
m^3/kg	cubic meter per kilogram
nm^3/kg	normal cubic meters per kilogram

11	**Heat**	
	kcal	kilo calories
	cal	calory
	mkcal	million kilo calories

12	**Gravity**	
	g	gravity constant : 9.81 metres per second per second

13	**Sp. heat**	
	Kcal/kg	kilo calories per kilogram
	Kcal/nm^3	kilo calories per normal cubic meter

14	**Heat consumption**	
	kcal/kg	kilo calories per kilogram
	kcal/nm^3	kilo calories per normal cubic meter
	mkcal/hr/m^2	million kilo calories per hour per square meter

15	**Rates of output, feed, flow**	
	tpd /TPD	tons per day
	tph	tons per hour
	tpa/ TPA	tons per annum
	MTPA/mtpa	million tons per annum
	m^3/sec	cubic meters per second
	m^3/min	cubic meters per minute
	m^3/hr	cubic meters per hour

16	**Power**	
	Kw	kilowatts
	p.f.	power factor
	kwh	kilo watt hour
	kw/t	kilowatts per ton
	kw/m	kilowatts per metre

17   **Specific output**

tpd/m^3 tons per day per cubic meter

18   **Cloth area**

m^3/min/m^2 cubic meters per minute per square meter

19   **Specific collection area**

m^2/m^3/sec   square meters per cubic meter per second

Section 1 BASICS

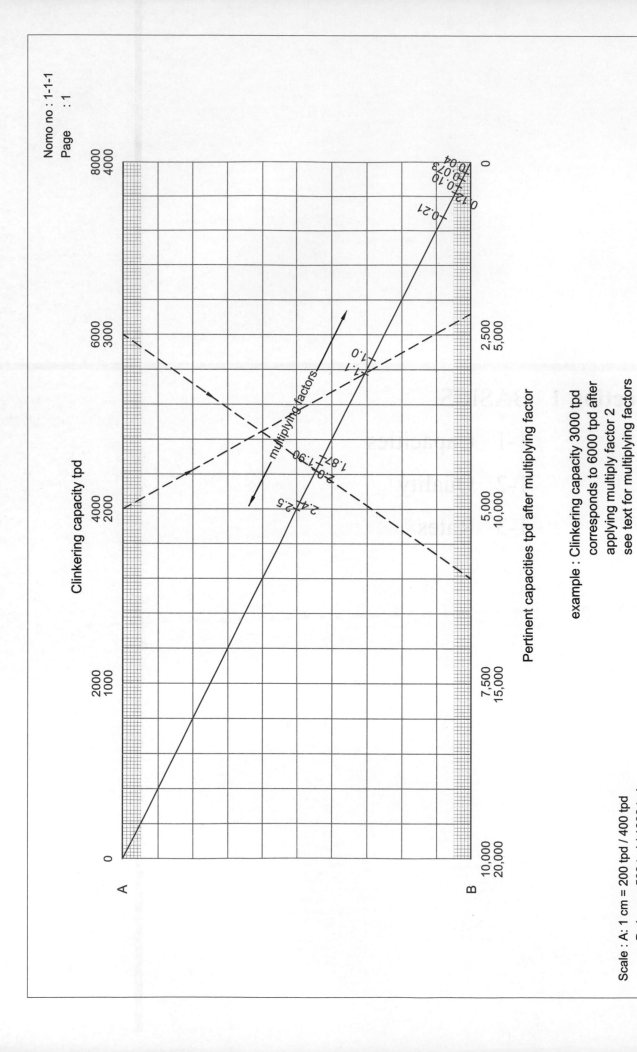

Clinkering capacity tpd

multiplying factors

Pertinent capacities tpd after multiplying factor

example : Clinkering capacity 3000 tpd
corresponds to 6000 tpd after
applying multiply factor 2
see text for multiplying factors

Nomogram for arriving at various capacities from given clinkering capacities in tpd

Scale : A: 1 cm = 200 tpd / 400 tpd
        B: 1 cm = 500 tpd / 1000 tpd

DEOLALKAR CONSULTANTS

Nomo no.:	1-1-1 page 2
Title:	Nomogram for arriving at various capacities from given Clinkering Capacities after pertinent multiplying factors
Useful for:	Working out hourly capacities and for storages and material handling equipment needed
Inputs:	clinkering capacities tpd multiplying factors
Output:	pertinent capacity in tpd See note below on multiplying factors
Scale:	A : 1 cm = 200/400 tpd clinkering capacity A : 1 cm = 500/1000 tpd pertinent capacity (same nomogram can be used for double the capacity)
How to use:	Various multiplying factors are as follows for

1	raisings in quarry	2.4
2	limestone to be crushed	1.9 / 2
3	clay, wet	0.21
4	sand	0.12
5	iron ore	0.04
6	coal wet	0.21-0.22
7	clinker design margin	1.1
8	gypsum wet	1.073
9	cement	1.27
10	raw meal	1.72
11	kiln feed	1.87
12	pulverised coal	0.20

Draw a line from given clinkering capacity on Line A through pertinent multiplying factor on Line 0-0 to meet line B. Read pertinent capacity on it.

Example:

clinkering capacity 3000 tpd (line A)
Multiplying factor say 2.0 (limestone crushed)
Daily requirement: 6000 tpd

Source	Handbook for Designing Cement Plants by S.P.Deolalkar

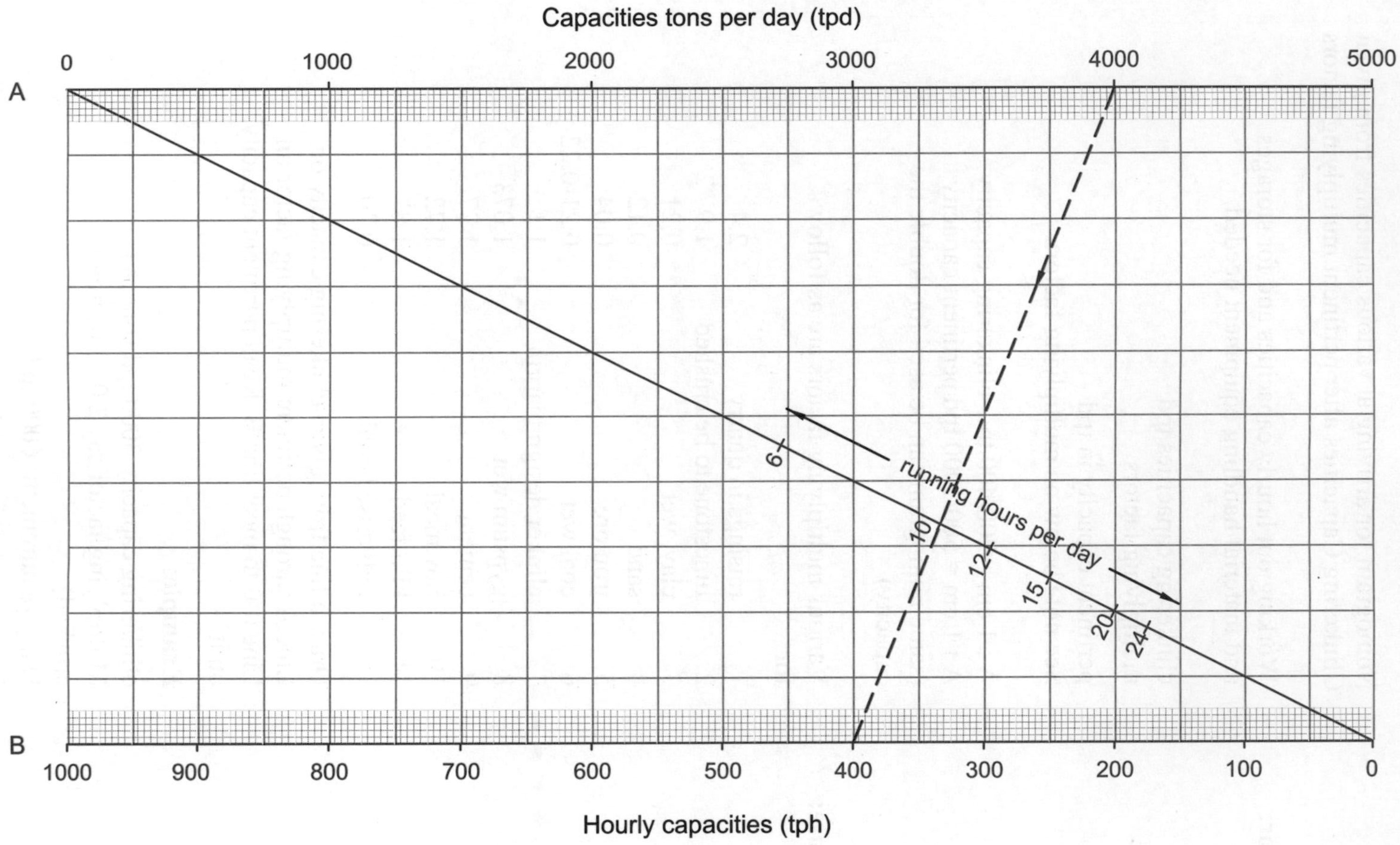

Nomo no : 1-1-2
Page : 1

Capacities tons per day (tpd)

running hours per day

24
20
15
12
10
9

Hourly capacities (tph)

example : Given capacity : 4000 tpd
no. of running hr. : 10
hourly capacity : 400 tph

Scale : A: 1 cm = 250 tpd
B: 1 cm = 50 tph

Nomogram for arriving at hourly capacities from daily capacities

DEOLALKAR CONSULTANTS

Nomo no.:	1-1-2 page 2
Title:	Nomogram for arriving at Hourly Capacities from Daily Capacities
Useful for:	quickly arriving at hourly capacities for further calculations
Inputs:	capacities in tons per day (tpd) running hours per day
Output:	hourly capacity (tph)
Scale:	A : 1 cm = 250 tpd B : 1 cm = 50 tph
How to use:	Draw a line from point showing given capacity in tpd on line A through given running hours on line 00 and extend it to meet line B. Read hourly capacity from the scale on it.

Example:

capacity tpd - 4000 (line A)
running hours - 10 (line 0-0)
hourly capacity – 400 tph

Source:	Handbook for Designing Cement Plants by S.P. Deolalkar

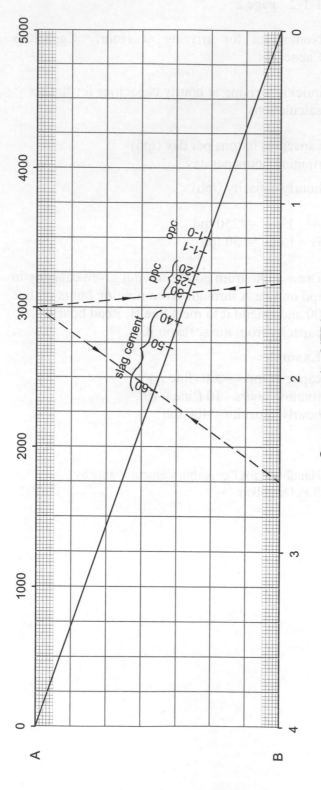

Clinkering capacity tpd

Cement capacity MTPA

example : clinkering capacity 3000 tpd corresponds to
a slag cement capacity of ~ 2.6 MTPA with 60% slag

Note : 1.0 - opc - 4% gypsum

1.1 - opc - 10% design margin

20-30% pozzolana in ppc – no design margin

40-60% slag in slag cement - no design margin

Nomogram for working out capacity in million tons per annum (MTPA)
from given clinkering capacity tpd for various types of cement

Scale : A: 1 cm = 250 tpd

B: 1 cm = 0.20 MTPA

DEOLALKAR CONSULTANTS

Nomo no.:	1-1-3 page 2
Title:	Nomogram for working out Capacity in Million Tons Per Annum (MTPA) from given clinkering capacity in tpd for various types of Cement
Useful for:	Quick reckoning of annual capacity in MTPA for OPC and Blended cements with different proportions of blending materials.
Inputs:	1 clinkering capacity in tpd 2 design margin 10 % 3 gypsum 4% 4 no. of working days per annum 330 5 proportions of blending materials like pozzolona and slag
Outputs:	Annual capacities of OPC, PPC and Slag Cements in various proportions of pozzolana and slag
Scale:	A : 1 cm = 250 tpd B : 1 cm = 0.2 MTPA
How to use:	Draw a line from given clinkering capacity on line A through pertinent marking on line 0-0 to meet line B and read annual capacity in MTPA

Note: on markings on line 0-0
1   OPC with 4 % gypsum
1.1   OPC with 10 % design margin
20-30 %   proportions of pozzolona in PPC
40-60 %   proportions of slag in Slag Cement

Example:

clinkering capacity -3000 tpd (line A)
slag cement with- 60 % slag (line 0-0)
slag cement capacity mtpa − 2.6

Source:	Handbook for Designing Cement Plants by S.P.Deolalkar

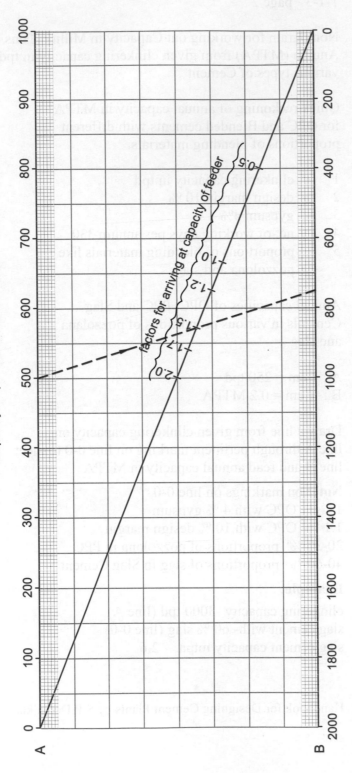

Capacity in tph of material to be processed

Capacity of corresponding feeder/conveyor in tph

factors for arriving at capacity of feeder

example : Machine rated capacity of 500 tph
will need a feeder of 750 tph with multiplying factor of 1.5
(for factors see text)

Nomogram for arriving at capacities of corresponding feeder/conveyors

Scale : A: 1 cm = 50 tpd
        B: 1 cm = 100 tph

DEOLALKAR CONSULTANTS

Nomo no.:	1-1-4 page 2
Title:	Nomogram for arriving at Capacities of corresponding Feeders and Conveyors to be used in conjunction with Nomo nos : 1-1-1 & 1-1-2
Useful for:	Arriving at capacities of feeders and conveyors in various sections after allowing margins usually provided in design of materials handling equipment
Inputs:	1 Capacities of main machines in tph 2 Margins to be provided in design of corresponding feeders and conveyors
Output:	Capacities of corresponding feeders and conveyors
Scale:	A : 1 cm = 50 tph B : 1 cm = 100 tph
How to use:	Margins and multiplying factors to be provided for main machines in different sections are as follows:

1	feeder to crusher	1.2
2	belt under crusher	1.5
3	most other feeders	1.2 – 1.3
4	main clinker conveyor	1.7
5	spillage conveyor for clinker	0.52
6	belt conveyors in packing plant	1.5
7	feeders for raw meal, coal	1.2

Draw a line from point showing rated capacity of main machine on line A and extend it to through desired multiplying factor on line 0 -0 to meet line B. Read required capacity of feeder/conveyor

Example:

capacity in tph: 500 (line A)
factor: 1.5 (line 0-0)
capacity of feeder: 750 tph (line B)

Source:	Handbook for Designing Cement Plants by S.P.Deolalkar

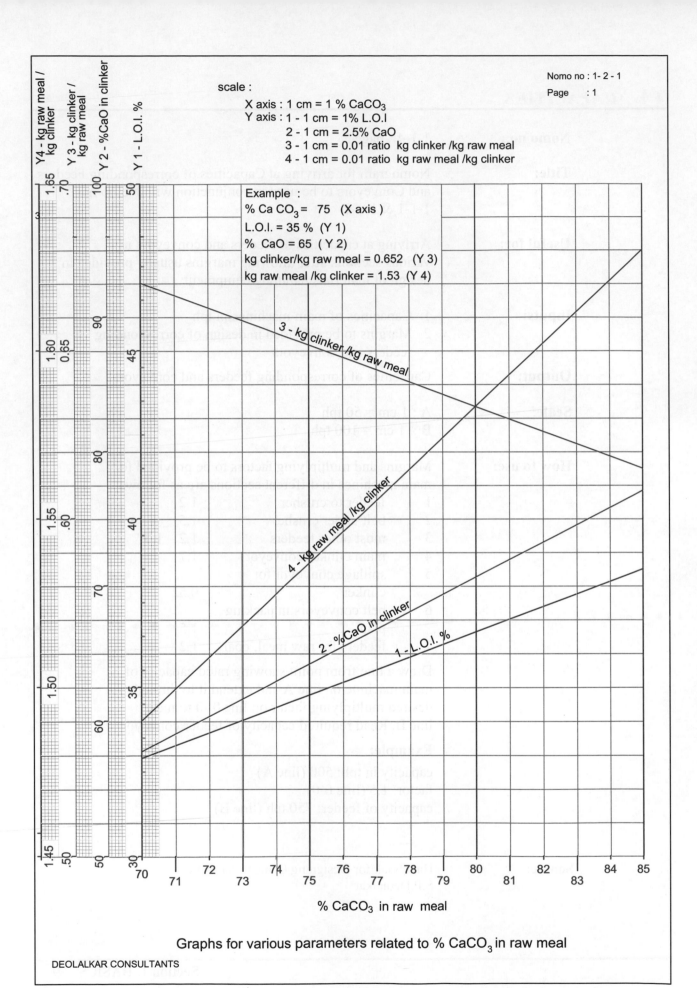

Graphs for various parameters related to % CaCO₃ in raw meal

DEOLALKAR CONSULTANTS

Nomo no.:	1-2-1 page no. 2
Title:	Graphs for various parameters related to % Ca CO_3 in raw meal
Useful for:	readily obtaining various quantities from known quantity of % carbonates in raw mix
Inputs:	1 % Ca CO_3 in raw meal
Outputs:	1 L.O.I. (loss on ignition)
	2 CaO in clinker
	3 Kg clinker/ kg raw meal
	4 Kg raw meal / kg clinker

Basis:

1   L.O.I. = % Ca CO_3 in raw meal $\times$ 0.44/100 (CO_2) + (100 – % Ca CO_3) $\times$ 0.07/100 (water in clay)

2   kg clinker /kg raw meal = 1- L.O.I.

3   kg raw meal/kg clinker = 1/ (1-L.O.I.)

4   % CaO in clinker = (% Ca $CO_3 \times$ 56/(1-L.O.I..)) $\times$ 100

Scale:

X axis : 1 cm = 1 % Ca CO_3

Y axis :   1   1 cm = 1 % L.O.I.

2   1 cm = 2.5 % CaO in clinker

3   1 cm = 0.01 kg clinker /kg raw meal

4   1 cm = 0.01 kg raw meal/kg clinker

How to use:

For a given % of $CaCO_3$ in raw meal on x axis read various values on corresponding scales on y axis

Example:

$CaCO_3$ = 75 %

L.O.I = 35 %

% CaO in clinker = 64.5 %

kg clinker/kg raw meal = 0.65

kg raw meal/kg clinker = 1/0.65 = ~1.53

Source: base data and formulae from Otto Labahn

Total % carbonates

| A | 50 | 55 | 60 | 65 | 70 | 75 | 80 | 85 | 90 | 95 | |

Loss on ignition %

| B | 25.5 | 27.4 | 29.2 | 31.1 | 32.9 | 34.6 | 36.6 | 38.5 | 40.3 | 42.2 | 44 |

kg Clinker/kg raw meal

| C | 0.75 | 0.73 | 0.71 | 0.69 | 0.67 | 0.65 | 0.63 | 0.62 | 0.60 | 0.58 | 0.56 |

Kg raw meal/Kg clinker

| D | 1.34 | 1.38 | 1.41 | 1.45 | 1.49 | 1.53 | 1.58 | 1.62 | 1.68 | 1.73 | 1.79 |

CaO in clinker

| E | 37.6 | 42.4 | 47.5 | 52.8 | 58.4 | 64.4 | 70.7 | 77.3 | 84.4 | 92.0 | 100 |

Scale: A - 2 cm = 5% carbonates
 B - = 1.8 -1.9% choose appropriate
 C - = 0.019 - 0.020 appropriate
 D - = 0.04 - 0.05 appropriate
 E - = 6-8% appropriate

example : total carbonate = 75% line A
 loss on ignition ~ 34.6% line B
 ratio raw meal to clinker ~ 0.65
 % in CaO in clinker ~ 64.4%

Nomogram for relations between % carbonates, Loss of Ignition,
% CaO in clinker & ratios kg clinker/kg raw meal & vice versa.

DEOLALKAR CONSULTANTS

Nomo no:	1-2-2 page 2
Title:	Nomogram for relations between % carbonates, loss on ignition, % CaO in clinker and ratios kg clinker/kg raw meal and vice versa
Useful for:	process calculations
Inputs:	1 % total carbonates in raw meal 2 7 % moisture in clay content
Outputs:	1 L.O.I. (loss on ignition) % 2 ratio of kg clinker/ kg raw meal 3 ratio kg raw meal/ kg clinker 4 % CaO in clinker

Formulae used:

let C = % total carbonates, then

$$L.O.I. = 0.44 \times C - \frac{(1-C)}{100} \times 7$$

Ratio clinker/raw meal = (1 − L.O.I./100) = a
Ratio raw meal/clinker = reciprocal of above = 1/a
% CaO in clinker = C × 56/a × 100

Scale:

A : 2 cm = 5 %
B : 2 cm = ~ 1.8-1.9 %
C : 2 cm = ~ 0.019 − 0.02 ratio
D : 2 cm = ~ 0.04 − 0.05 ratio
E : 2 cm = ~ 6-8 %
~ values are on account of rounding off.
Divide pertinent 2 cm in equal parts

How to use:

For a given carbonate content on line A, draw a line vertically to meet lines B, C, D , E and read respective values of L.O.I etc on them

Example:

75 % Carbonate (line A) would mean 34.8 % L.O.I. (line B) and 0.65 clinker/meal ratio (line C),
1.53 ratio raw meal to clinker (line D) and
64.4 % CaO in clinker (line E)

Source: constructed
formulae from Otto Labahn

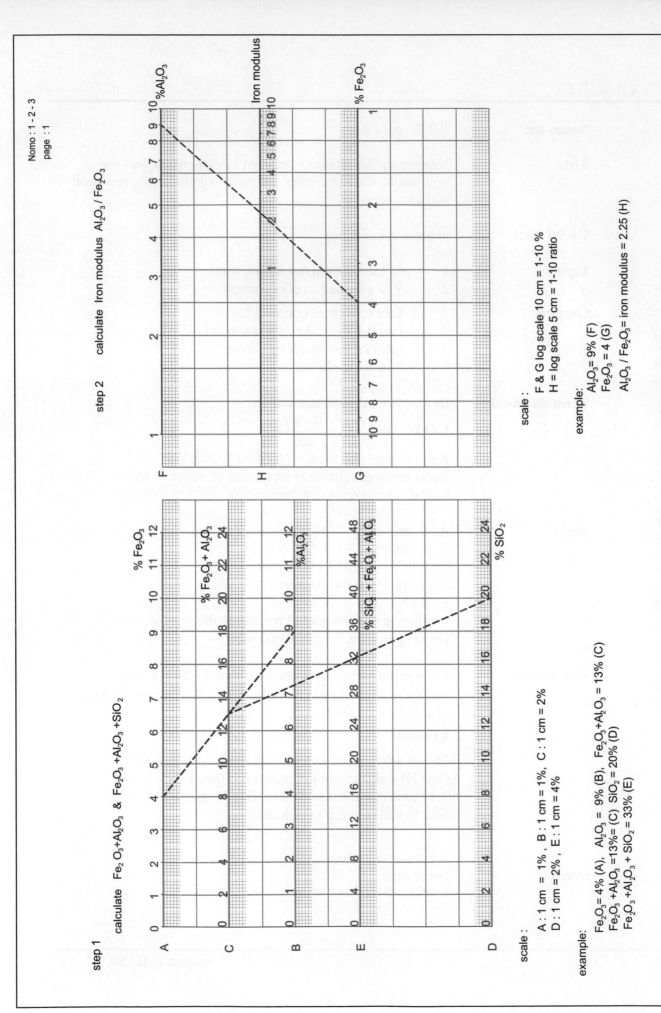

step 2   calculate Iron modulus Al_2O_3 / Fe_2O_3

10 %Al_2O_3

Iron modulus

% Fe_2O_3

F   H   G

scale :

F & G log scale 10 cm = 1-10 %
H = log scale 5 cm = 1-10 ratio

example:

Al_2O_3 = 9% (F)
Fe_2O_3 = 4 (G)
Al_2O_3 / Fe_2O_3 = iron modulus = 2.25 (H)

step 1   calculate $Fe_2O_3 + Al_2O_3$ & $Fe_2O_3 + Al_2O_3 + SiO_2$

% Fe_2O_3

% $Fe_2O_3 + Al_2O_3$

%Al_2O_3

% $SiO_2 + Fe_2O_3 + Al_2O_3$

% SiO_2

A   C   B   E   D

scale :

A : 1 cm = 1% , B : 1 cm = 1%, C : 1 cm = 2%
D : 1 cm = 2% , E : 1 cm = 4%

example:

Fe_2O_3 = 4% (A), Al_2O_3 = 9% (B), $Fe_2O_3 + Al_2O_3$ = 13% (C)
$Fe_2O_3 + Al_2O_3$ =13%= (C) SiO_2 = 20% (D)
$Fe_2O_3 + Al_2O_3 + SiO_2$ = 33% (E)

Nomogram for calculating various modulii from detailed
chemical analysis of raw materials on oxide basis

DEOLALKAR CONSULTANTS

Nomo no.:	1-2-3 page 2
Title:	Nomogram for calculating various modulii from detailed chemical analysis of raw materials on oxide basis
Useful for:	quickly determining suitability of making cement and for assessing requirements of correcting materials
Inputs:	1 Detailed chemical analysis of raw materials on oxide basis % SiO_2 Al_2O_3 Fe_2O_3 CaO MgO 2 Desired values of : 1 Hydraulic modulus 2 Silica ratio 3 Iron modulus

Definitions:

Hydraulic modulus : $CaO / (SiO_2 + Al_2O_3 + Fe_2O_3)$
Usual range : 1.7-2.2

Silica modulus : $SiO_2 / (Al_2O_3 \; Fe_2O_3)$
Usual range : 1.2-4.0

Iron modulus : Al_2O_3 / Fe_2O_3
Usual range : 1-4

Outputs: modulii as explained above

Scales:

A	: 1 cm = 1 %
B	: 1 cm = 1 %
C	: 1 cm = 2 %
D	: 1 cm = 2 %
E	: 1 cm = 4 %
F & G	: log scales : 10 cms = 1-10 %
H	: log scale : 5 cms = 1-10 ratio
I & J	: log scales : 10 cms = 10-100 %
K	: log scale : 5 cms = 1-10 ratio
L & M	: log scale : 20 cms = 10-100 %
N	: log scale : 10 cms = 1-10 ratio

How to use: **Example:**

detailed chemical analysis of raw mix :
% SiO_2 Al_2O_3 Fe_2O_3 CaO + MgO
20   9   4   66

step 1 : Find Al_2O_3 Fe_2O_3
draw line from 4 (A) to 9 (B) and read sum 13 (C)

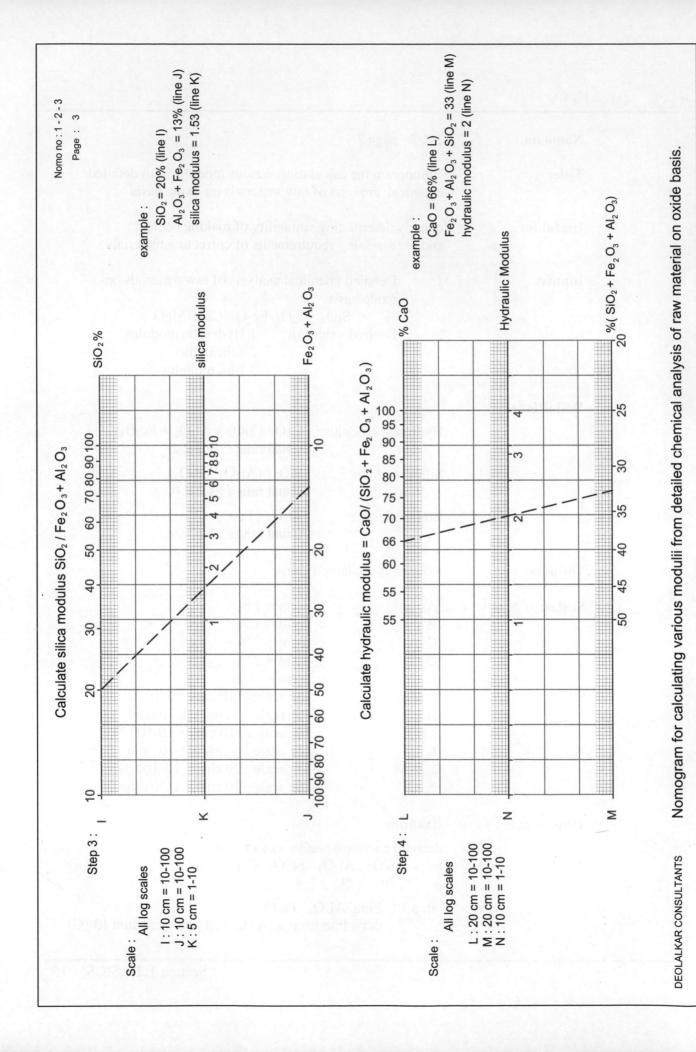

Nomo no : 1 - 2 - 3
Page : 3

Calculate silica modulus SiO₂ / Fe₂O₃ + Al₂O₃

example :

SiO₂ = 20% (line I)
Al₂O₃ + Fe₂O₃ = 13% (line J)
silica modulus = 1.53 (line K)

SiO₂ %

silica modulus

Fe₂O₃ + Al₂O₃

Step 3 : I

K

J

Scale :

All log scales

I : 10 cm = 10-100
J : 10 cm = 10-100
K : 5 cm = 1-10

Calculate hydraulic modulus = CaO/ (SiO₂ + Fe₂O₃ + Al₂O₃)

example :

CaO = 66% (line L)
Fe₂O₃ + Al₂O₃ + SiO₂ = 33 (line M)
hydraulic modulus = 2 (line N)

% CaO

Hydraulic Modulus

%(SiO₂ + Fe₂O₃ + Al₂O₃)

Step 4 : L

N

M

Scale :

All log scales

L : 20 cm = 10-100
M : 20 cm = 10-100
N : 10 cm = 1-10

DEOLALKAR CONSULTANTS   Nomogram for calculating various modulii from detailed chemical analysis of raw material on oxide basis.

Nomo no.:   1-2-3   page 4

step 2 : Find SiO_2 Al_2O_3 Fe_2O_3
draw line from 13 (C) to 20 (D) and read sum 33 (E)

step 3 : Find iron modulus Al_2O_3 / Fe_2O_3
draw a line from 9 (F) to 4 (G) and read Iron
modulus 2.25 (H)

step 4 : Find silica modulus SiO_2 / (Al_2O_3 Fe_2O_3)
draw a line from 20 (I) to 13 (J) and read Silica
modulus 1.53 (K)

step 5 : Hydraulic modulus CaO / (SiO_2 + Al_2O_3 + Fe_2O_3)
draw a line from 66 (L) to 33 (M) and read
Hydraulic modulus 2 (N)

Source: formulae for moduli from Otto Labahn
nomograms constructed

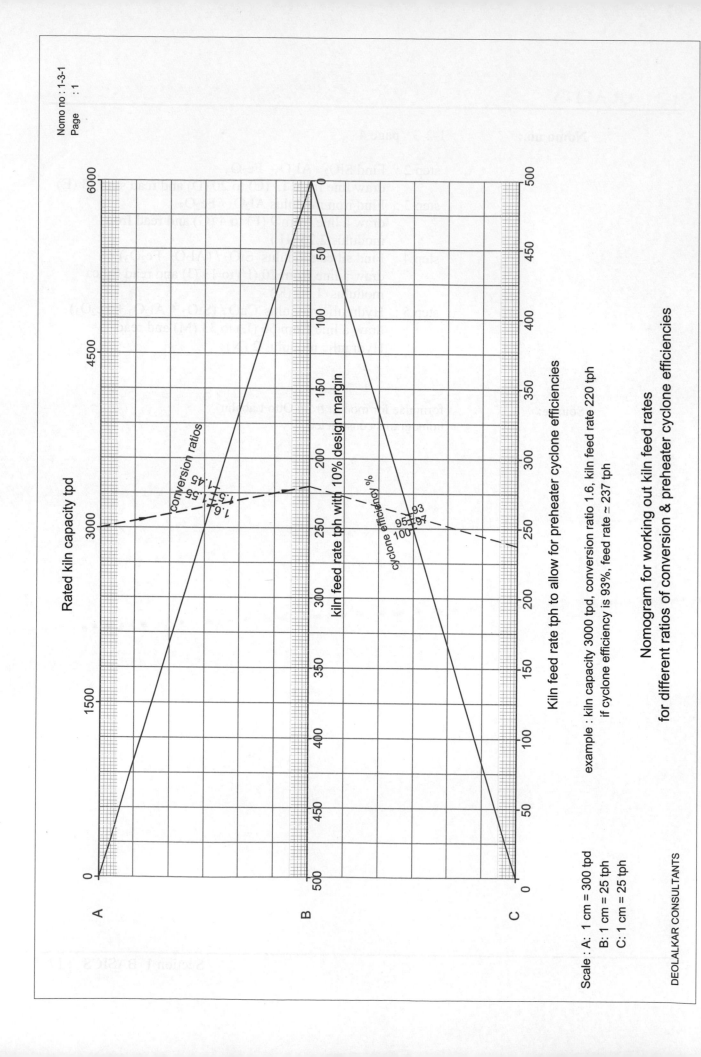

Rated kiln capacity tpd

Kiln feed rate tph to allow for preheater cyclone efficiencies

example : kiln capacity 3000 tpd, conversion ratio 1.6, kiln feed rate 220 tph
if cyclone efficiency is 93%, feed rate ≈ 237 tph

Nomogram for working out kiln feed rates
for different ratios of conversion & preheater cyclone efficiencies

Scale : A: 1 cm = 300 tpd
        B: 1 cm = 25 tph
        C: 1 cm = 25 tph

DEOLALKAR CONSULTANTS

Nomo no.:	1-3-1 page 2
Title:	Nomogram for working out kiln feed rates for different ratios of conversion and different preheater cyclone efficiencies
Useful for:	Finding out kiln feed rates for different capacities of kiln

Inputs:

1   rated capacities of kiln in tpd
2   conversion ratios raw meal/clinker
3   efficiencies of preheater cyclones in %

Outputs:

1   kiln feed rates with design margin of 10 %
     in tph for different ratios of conversion
2   kiln feed rates to allow for efficiencies of
     preheater cyclones in tph

Scale:

A : 1 cm = 300 tpd
B : 1 cm = 25 tph
C : 1 cm = 25 tph

How to use:

Draw a line from rated kiln capacity on line A through applicable ratio of conversion on line 0-0 to meet line B. Read raw meal in tph with 10 % design margin.
From this point draw a line through pertinent cyclone efficiency on line 0-0' to meet line C and read kiln feed rate in tph.

Example:

kiln capacity – 3000 tpd (line A)
conversion ratio – 1.6 (line 0-0)
kiln feed rate - 220 tph (line B)
cyclone efficiency - 93 % (line 0-0')
kiln feed rate after allowing for
cyclone efficiency - ~237 tph

Source: constructed

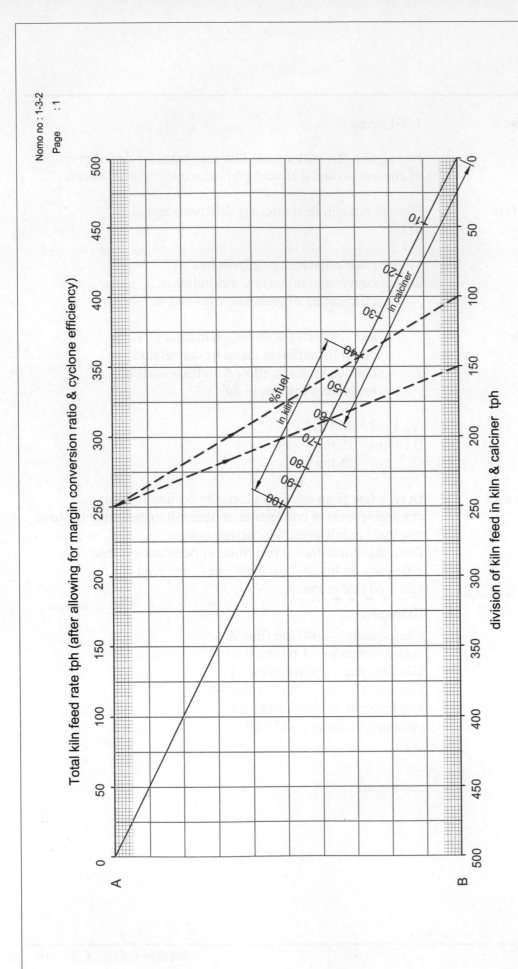

Total kiln feed rate tph (after allowing for margin conversion ratio & cyclone efficiency)

division of kiln feed in kiln & calciner tph

% fuel

in kiln

in calciner

example - feed rate to kiln 60%;   feed to calciner 40%   (line 0-0)
total feed rate = 250 tph   (line A)
feed rate to kiln = 150 tph and to calciner = 100 tph   (line B)

Nomogram for working out division of feed rates in kiln & calciner

Scale : A: 1 cm = 25 tph
        B: 1 cm = 25 tph

DEOLALKAR CONSULTANTS

Nomo no.:	1-3-2 page 2
Title:	Nomogram for working out division of feed rates in kiln and calciner
Useful for:	for designing feed systems for kiln and calciner
Inputs:	1. Total feed rates in tph taking into account design margin, conversion ratio and cyclone efficiency see nomo no : 1-3-1 2. % fuels in kiln and calciner
Output:	feed rates in tph in kiln and calciner
Scale:	A : 1 cm = 25 tph B : 1 cm = 25 tph
How to use:	Draw a line from total feed rate on line A, through % fuel in kiln on line 0-0 and extend to meet line B and read feed rate in kiln in tph

Example:

total feed rate – 250 tph (line A)
fuel in kiln 40 % - (line 0-0)
feed rate in kiln – 100 tph (line B)
fuel in calciner 60 % - (line 0-0)
feed rate in calciner - 150 tph (line B)

Source:	constructed

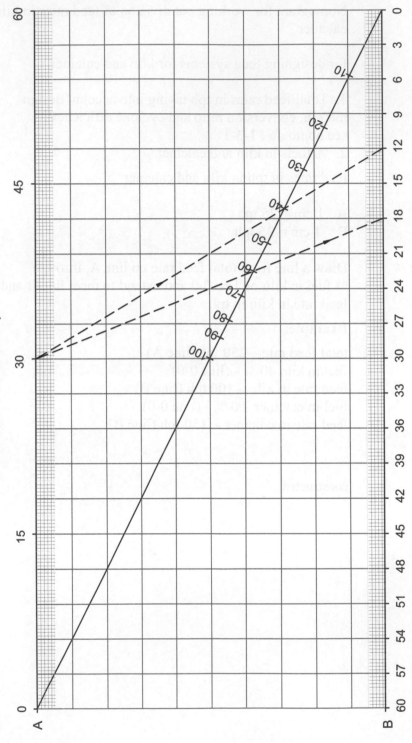

Total rate of fuel fired in tph

Fuel fired in kiln /calciner tph

example : from point 30 tph on line A, draw line through 40% fuel to meet line B to read
fuel in kiln - 12 tph; draw a line through 60% and read fuel fired in calciner 18 tph

Nomogram for working out fuel fired in kiln & calciner in tph respectively from total fuel fired

Scale : A: 1 cm = 3 tph
        B: 1 cm = 3 tph

DEOLALKAR CONSULTANTS

Nomo no.:	1-3-3 page 2
Title:	Nomogram for working out fuel fired in kiln and calciner in tph respectively from total fuel fired
Useful for:	knowing directly in tph quantities of fuel fired in kiln and calciner separately
Inputs:	1. total fuel fired in tph 2. % fuel fired in kiln and calciner respectively
Output:	rates of fuel fired in tph in kiln and calciner
Scale:	A : 1cm = 3 tph B : 1 cm = 3 tph
How to use:	draw a line from given total rate of fuel fired on line A through % fuel fired in kiln or calciner on line 0-0 and extend to meet line B and read quantity of fuel fired in kiln or calciner as the case may be.

Example:

Total fuel: 30 tph (line A)
60 % in calciner = 18 tph (line 0-0 & line B)
40 % in kiln = 12 tph (line 0-0 & line B)

Source:	constructed

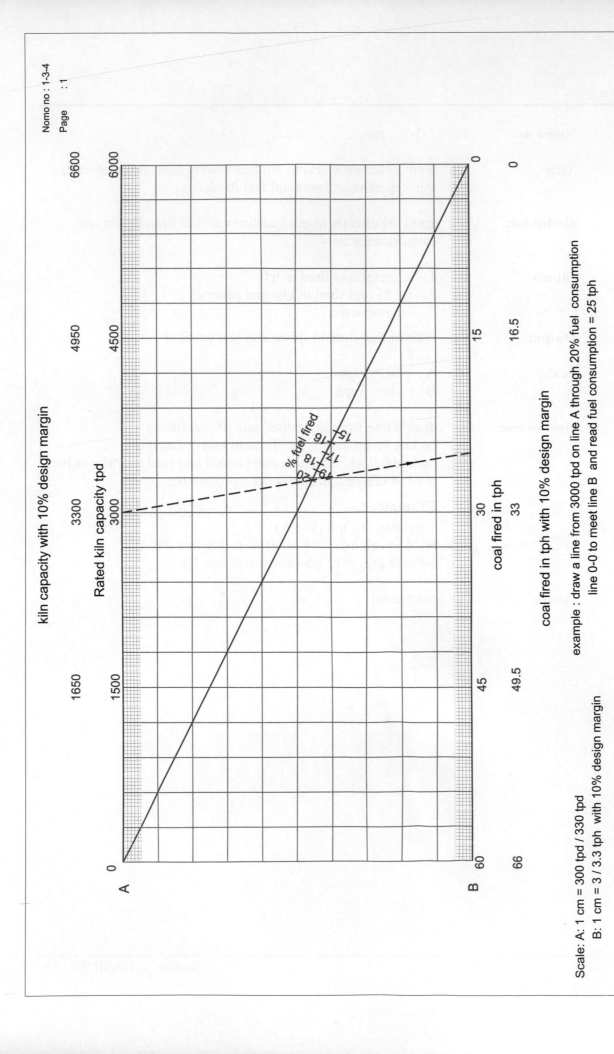

Nomo no : 1-3-4
Page : 1

kiln capacity with 10% design margin

Rated kiln capacity tpd

% fuel fired

coal fired in tph

coal fired in tph with 10% design margin

example : draw a line from 3000 tpd on line A through 20% fuel consumption
line 0-0 to meet line B and read fuel consumption = 25 tph

Nomogram for working out coal consumption in tph
for different kiln capacities & rates of fuel consumption

Scale: A: 1 cm = 300 tpd / 330 tpd
       B: 1 cm = 3 / 3.3 tph with 10% design margin

DEOLALKAR CONSULTANTS

Nomo no.:	1-3-4 page 2
Title:	Nomogram for working out coal consumption in tph for different kiln capacities and rates of fuel consumption
Useful for:	working out capacities of coal grinding and coal firing systems
Inputs:	1 kiln capacity tpd 2 sp. Fuel consumption expressed as % of clinker
Output:	consumption of coal in tph
Scale:	A : 1 cm = 300/330 tpd B: 1 cm = 3/3.3 tph
How to use:	Draw a line from given kiln capacity on line A through sp. Fuel consumption on line 0-0 to meet line B and read coal consumption in tph. Read appropriate scales for capacities with design margin

Example:

kiln capacity – 3000 tpd (line A)
Sp. fuel consumption - 20 %   (line 0-0)
Coal firing rate tph – 25 tph (line B)

Source:	constructed

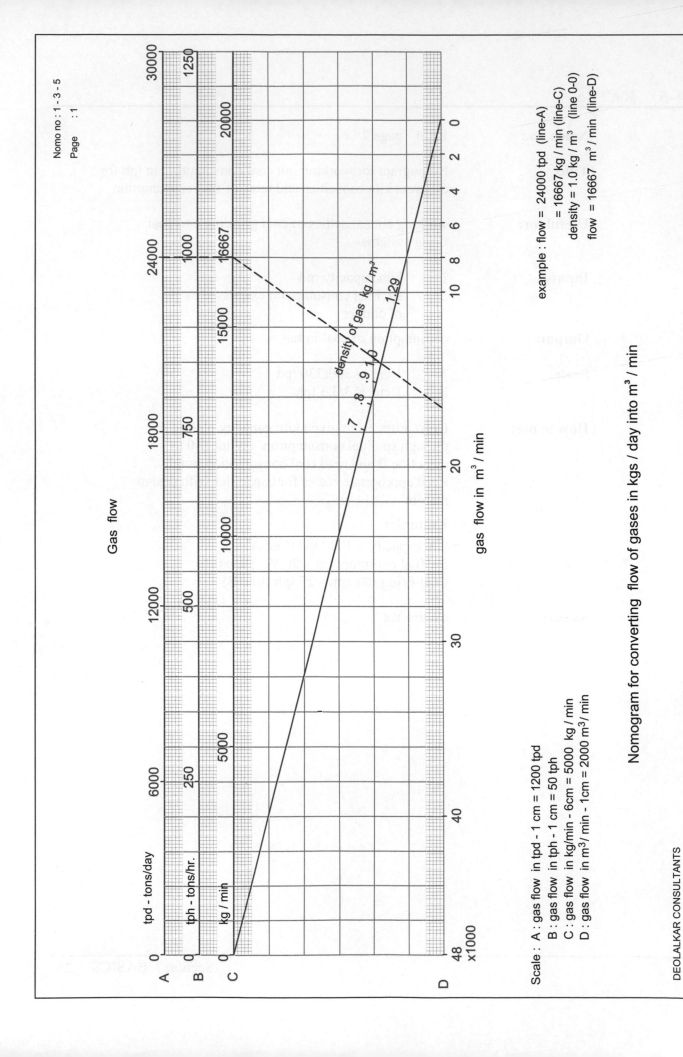

Gas flow

A — tpd - tons/day

B — tph - tons/hr.

C — kg / min

density of gas kg / m³

gas flow in m³ / min

example : flow = 24000 tpd (line-A)
         = 16667 kg / min (line-C)
density = 1.0 kg / m³ (line 0-0)
flow = 16667 m³/ min (line-D)

Scale : A : gas flow in tpd - 1 cm = 1200 tpd
        B : gas flow in tph - 1 cm = 50 tph
        C : gas flow in kg/min - 6cm = 5000 kg / min
        D : gas flow in m³/ min - 1cm = 2000 m³/ min

Nomogram for converting flow of gases in kgs / day into m³ / min

DEOLALKAR CONSULTANTS

Nomo no.:	1-3-5 page 2
Title:	Nomogram for converting flow of gases in kg / day into m^3/ min
Useful for:	process design calculations
Inputs:	1. gas flow in kg/day 2. density of gas in kg/m^3
Output:	gas flow in m^3/min
Scale:	A : 1 cm = 1200 tpd B : 1 cm = 50 tph C : 6 cm = 5000 kg/min D : 1 cm = 2000 m^3/min
How to use:	from line A (tpd), draw line at right angle to lines B and C to know quantities in tph and kgs / min. From point on line C (kgs /min) draw line through given density on line 0-0 and read gas flow in m^3/min on line D

Example:

Flow : 24000 tpd
$\qquad$ = 1000 tph
$\qquad$ = 16667 kg/min
density = 1.0 kg / m^3
∴ flow = 16667 m^3/min

Source:	constructed

Section 2 PHYSICAL PROPERTIES

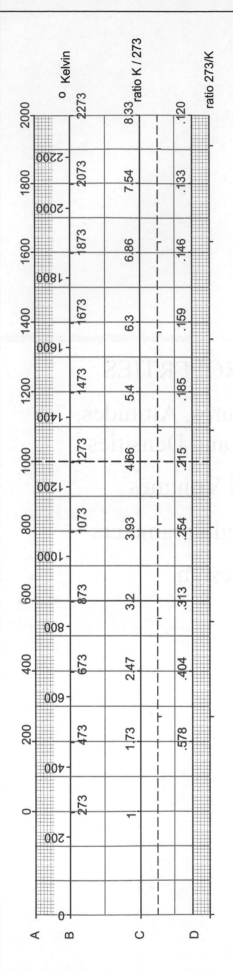

Temperature °C

Example : temp. - 1000 °C (line A)
         O Kelvin = 1273 (line B)
         ratio K / 273 = 4.66 (line C)
         ratio 273/K = 0.215 (lineD)

Nomogram for converting temperature in °C to Kelvin
and for working out ratios K / 273 & 273 / K

Scale : A : 1 cm = 100 °C
        B : 1 cm = 100 °K
        C : K / 273 marked up scale
        D: 273/ K marked up scale

DEOLALKAR CONSULTANTS

Nomo no.:	2-1-1 page 2
Title:	Nomogram for converting temperature in °C to Kelvin and for working out ratios of K/273 and 273/K
Useful for:	converting m^3 to nm^3 and vice versa
Input:	temperature in °C
Outputs:	1 temperature in Kelvin K 2 ratio K/273 3 ratio 273/K
Scale:	A : 1 cm = 100 °C B : 1 cm = 100 °K C : marked up scale K/273 D : marked up scale 273/K
How to use:	Draw a perpendicular line from the given temp. t on line A to line B and read K for obtaining ratio K/273, extend it to meet line C and read ratio on it. For obtaining ratio 273/K, extend the line to meet line D and read ratio on it. **Example:** Temp t = 1000 °C (line A) Temp. Kelvin K = 1273 (line B) Ratio K/273 = ~ 4.66 (line C) Ratio 273/K = ~ 0.215 (line D)
Source:	constructed

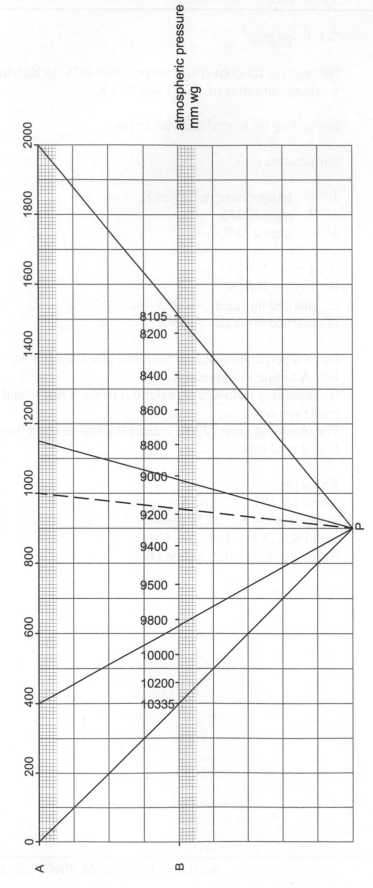

atmospheric pressure
mm wg

Altitude in metres

8105
8200

8400

8600

8800

9000

9200

9400

9500

9800

10000

10200

10335

2000 1800 1600 1400 1200 1000 800 600 400 200 0

A

B

P

example: altitude = 1000 m (line A)
             atmospheric pr. = 9180 mm (line B)

scale: A : 1 cm = 100 m
          B : specially constructed scale
          P : point of reference

Nomogram for obtaining atmospheric pressure at different altitudes

DEOLALKAR CONSULTANTS

Nomo no.:	2-1-2 page 2
Title:	Nomogram for obtaining atmospheric pressures at different altitudes
Useful for:	system design of gas flows at different temperatures and pressures ; for converting nm^3 to m^3 and vice versa
Inputs:	1 altitude at site in m above sea level 2 atmospheric pressure at sea level = 10335 mm wg
Output:	atmospheric pressure at altitude
Scale:	A : 1 cm = 100 m B : specially constructed scale P : point of reference
How to use:	draw a line from P to join pt. of altitude on line A and read atmospheric pressure at that altitude on line B

Example:
altitude = 1000 m (line A)
atmospheric pressure = ~ 9180 mm wg (B)

Source:	constructed

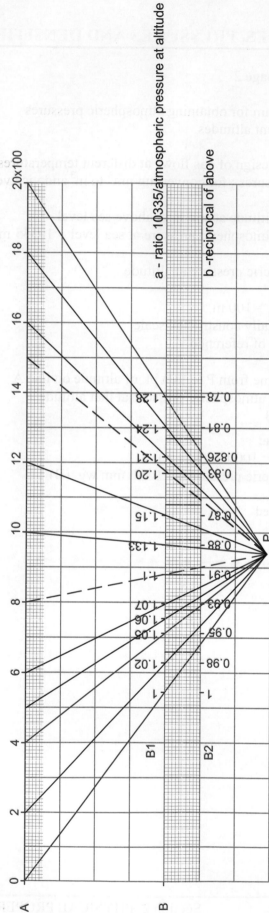

Altitude in metres

a - ratio 10335/atmospheric pressure at altitude

b - reciprocal of above

Example :

Altitude : 1500 m (line A)

Ratio a = 1.2 (line B1)

Ratio = 1/a = 0.83 (line B2)

scale: A : 1 cm = 100 metres

B1 : specially constructed scale

B2 : specially constructed scale

P : point of reference

Nomogram for working out ratios of atmospheric pressures at altitude and sea level
and reciprocal thereof

DEOLALKAR CONSULTANTS

Nomo no.:	2-1-3 page 2
Title:	Nomogram for working out ratios of atmospheric pressures at site altitude and sea level and reciprocal thereof and for obtaining ratio when system is under negative pressure
Useful for:	converting nm^3 to m^3 and vice versa and for working out specifications for fans in the system

Inputs:

1   altitudes in m above sea level
2   atmospheric pressures at altitude vide
    nomo no. : 2-1-2
3   draft in mm wg at inlet of fan
4   atmospheric pressure at sea level 10335 mmwg

Outputs:

1   ratio 10335/ atmospheric pr. at altitude **(a)**
2   ratio pr. at altitude /10335 **(b)**
3   ratio 10335 / (pr. at altitude – draft at fan inlet) **(c)**

Scale :

A : 1 cm = 100 m altitude
B1 : specially constructed scale for ratio **(a)** Altitude
B2 : **(b)** reciprocal of above
P : point of reference
C : 1 cm = 100 m
D : specially constructed scale for draft at fan inlet
E : scale B1 reversed to read ratio **(c)**

How to use:

1   ratio 10335/barometric pressure and vice versa (a and b)
    Draw a line from point of reference P to point of altitude on line A and read ratio **(a)** on line B1
2   for ratio **(b)** read on line B2
3   for ratio **(c)**
    draw a line from altitude on line C thru draft at fan inlet on line D and extend to meet line E and read ratio on it.

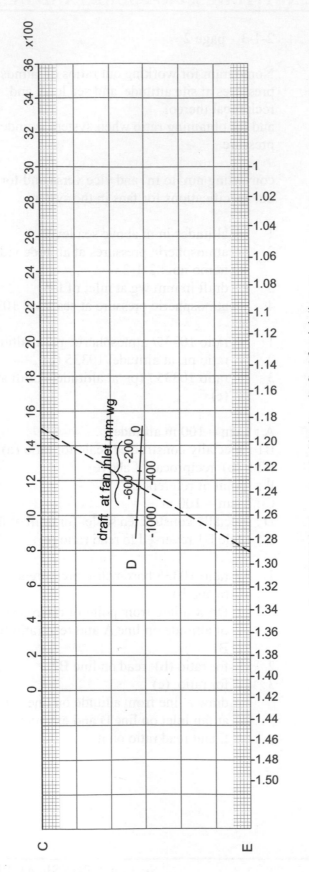

Altitude in metres

draft at fan inlet mm wg

ratio 10335 / (atmos. pr. at altitude - draft at fan inlet) = c

Nomogram for finding ratio 10335 / (atmos pr. at altitude - draft at fan inlet)

Example:

altitude = 1500 m (line C)

draft at fan inlet = -600 mm wg (line D)

Ratio c = ~ 1.29   (line E)

Scale : C : 1 cm = 200m

D : specially constructed

E : Log scale 100cm = 1-10

DEOLALKAR CONSULTANTS

2-1-3   page 4

Example :
1   altitude = 1500 m (line A)
         ratio **(a)** = 1.2 (line B1)
2   ratio **(b)** = 0.833 (line B2)
3   altitude = 1500 m (line C)
         draft at fan inlet = - 600 mm wg (line D)
         ratio **(c)** = 1.29 (line E)

Source: constructed

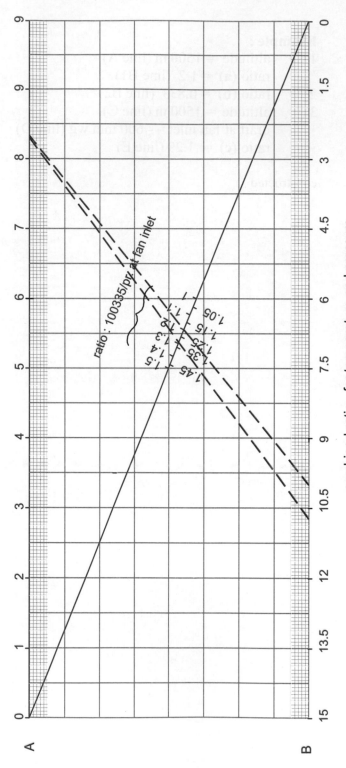

Ratio of Kelvin / 273

combined ratio - for temperature and pressure

ratio : 100335/pz at fan inlet

example :

ratio : K/273 = 8.33 corresponding to 2000° C (line A)

ratio 10335/pressure at altitude = 1.2 (line (0-0)

total ratio = 10 (line B)

Nomogram to arrive at total multiplier for both temp. and pressure to convert nm^3 to m^3 and vice versa

Scale : A : 1 cm = 0.5
        B : 1 cm = 0.75

DEOLALKAR CONSULTANTS

Nomo no.:	2-1-4 page 2
Title:	Nomogram to arrive at total multiplier for both temperature and pressure to convert nm^3 to m^3 and vice versa
Useful for:	working out actual gas volumes to be handled in a system and by fans in the system
Inputs:	1 ratio K/273 see nomo no 2-1-1
	2 ratio 10335/ barometric pr. at altitude see nomo no 2-1-3 page 1
	3 ratio 10335/barometric pr.at altitude- draft at fan inlet see nomo no 2-1-3 page 3
Output:	total multiplier
Scale:	A : 1 cm = 0.5
	B : 1 cm = 0.75
How to use:	Draw a line from ratio K/273 on line A through ratio 10335/ baro pr. at fan inlet on line 0-0 to meet line B and read total multiplier on it

Example:
ratio K/273 = 8.33 (corresponding to temp 2000 $^\circ$ C)
on line A
1. ratio 10335/atmos. pr at altitude 1500 m = 1.2
   from line B (nomo no 2-1-3 page 1) on line 0-0
   read total multiplier = ~ 10 on line B
   E (nomo no 2-1-3 page 3) on line 0-0
   read total ratio = ~ 10 on line B
2. draft at fan inlet –600 mm
   ∴ ratio 10335/pr.at 1500 m –600mm =1.29 from
   line E (nomo no 2-1-3 page 3)
   ∴ total multiplier = 8.33 × 1.29 ≃ 10.75 on line B

Source:	constructed

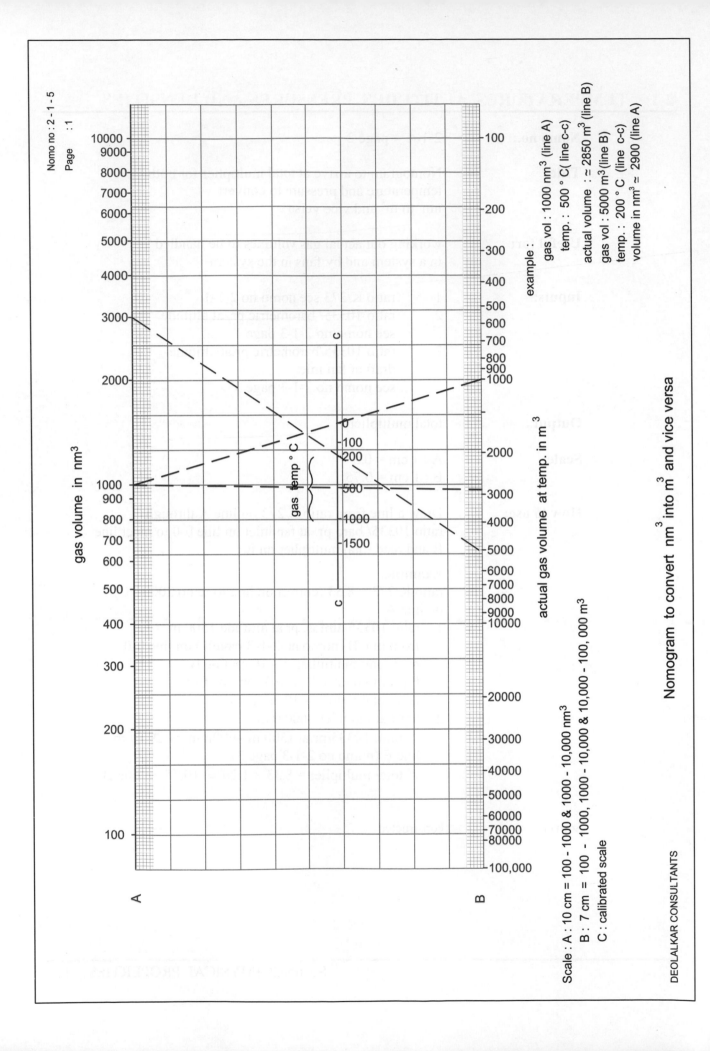

gas volume in nm³

gas temp ° C

actual gas volume at temp. in m³

example :

gas vol : 1000 nm³ (line A)
temp. : 500 ° C (line B)
actual volume : ≃ 2850 m³ (line B)
gas vol : 5000 m³ (line B)
temp. : 200 ° C (line c-c)
volume in nm³ ≃ 2900 (line A)

Scale : A : 10 cm = 100 - 1000 & 1000 - 10,000 nm³
B : 7 cm = 100 - 1000, 1000 - 10,000 & 10,000 - 100, 000 m³
C : calibrated scale

Nomogram to convert nm³ into m³ and vice versa

DEOLALKAR CONSULTANTS

Nomo no.:	2-1-5 page 2
Title:	Nomogram to convert nm^3 to m^3 and vice versa
Useful for:	sizing ductings in a system range 0-1500 $^\circ$C and 0-10000 nm^3
Inputs:	1 gas volume in nm^3 2 temperature in $^\circ$C 3 altitude assumed at sea level
Output:	Actual volume in m^3
Scale:	A : log scale -10 cm = 100-1000,1000-10000 B : log scale –7 cm = 100-1000 etc C : calibrated scale for temperatures 0-1500 $^\circ$C
How to use:	draw a line from given volume in nm^3 on line A thru temp. on line C to meet line B and read actual volume in m^3 Reverse procedure to convert m^3 into nm^3 **Example:** Gas volume - 1000 nm^3 (line A) Temperature - 500 $^\circ$C (line C) Actual volume - ~ 2850 m^3 (line B) Actual volume - 5000 m^3 (line B) Temperature - 200 $^\circ$C (line C) Volume in nm^3 ~ 2900 (line A)
Source:	constructed

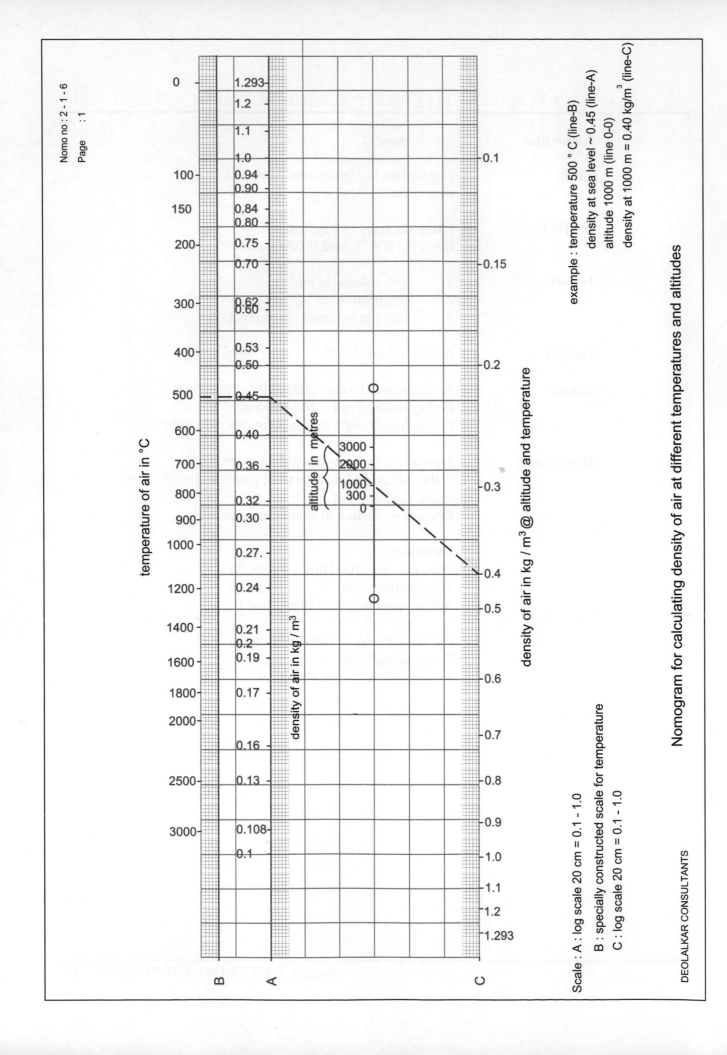

Nomo no : 2 - 1 - 6
Page   : 1

temperature of air in °C

density of air in kg / m³

altitude in metres

density of air in kg / m³ @ altitude and temperature

example : temperature 500 ° C (line-B)
          density at sea level ~ 0.45 (line-A)
          altitude 1000 m (line 0-0)
          density at 1000 m = 0.40 kg/m³ (line-C)

Scale : A : log scale 20 cm = 0.1 - 1.0
        B : specially constructed scale for temperature
        C : log scale 20 cm = 0.1 - 1.0

Nomogram for calculating density of air at different temperatures and altitudes

DEOLALKAR CONSULTANTS

Nomo no.:	2-1-6 page 2
Title:	Nomogram for calculating density of air at different altitudes and temperatures
Useful for:	many system design calculations ; for calculating velocity pressure etc.

Inputs:

1   temperature of air – 0-3000 ° C
2   altitude   - 0- 3000 m
3   density of air at n.t.p. - 1.293 kg/m^3

Output: density of air at different temperatures and Altitudes

Scale:

1   density of air at different temperatures at
     sea level
A : log scale : 20 cm = 0.1-1.0 density of air in kg / m^3
B : constructed scale for temperature 0-3000 ° C

2   density of air at different altitudes
C : log scale : 20 cm = 0.1-1.0 density of air in kg / m^3
O-O   specially constructed scale for altitude 0- 3000 m

How to use:

1   first determine density at sea level at different
     temperatures using lines A and B
2   from the pertinent point of density on line A
     draw a line through given altitude on line
     O-O and extend to meet line C and read density at
     that altitude on it in kg / m^3

Example:
Temperature of air 500 ° C (line B)
Corresponding density at sea level = ~ 0.45 kg / m^3
                                              (line A.)
Density at 1000 m altitude = ~ 0.4 kg/m^3 (line C)

Source: constructed

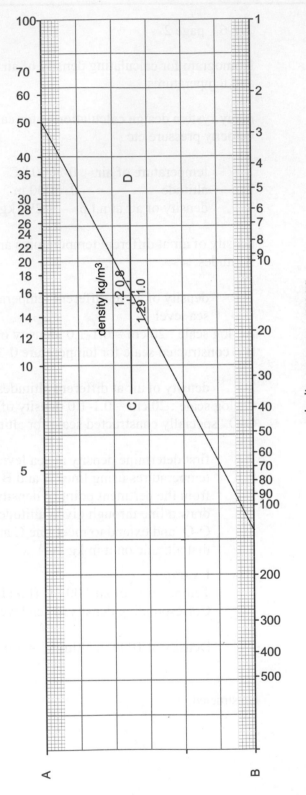

velocity m / sec

velocity pressure mmwg

Scale : A : log scale 10 cm = 10-100
        B : log scale 7 cm = 1-10, 10-100 etc

example: velocity = 50 m /sec (line A)
         density - 1 kg/m³ (line C-D)
         velocity pressure ≈ 127 mmwwg (line B)

Nomogram for working out velocity pressure from velocity & density of air / gas

DEOLALKAR CONSULTANTS

Nomo no.:	2-1-7 page 2
Title:	Nomogram for calculating velocity pressure from given velocity and density of air / gas
Useful for:	working out pressure loss in a system involving flow of gases etc.
Inputs:	1 velocity of air/gas in m/sec. 2 density of air/gas in kg/m^3
Output:	Velocity pressure in mm wg Velocity pressure $= (v^2/2g) \times$ density where v = velocity in m/sec. and g is gravity constant g = 9.81/m/sec/sec
Scale:	A : log scale 10 cm = 1-10,10-100 B : log scale 7 cm = 1-10, 10-100 C : constructed scale
How to use:	draw a line from given velocity 50 m/sec (line A) thru density 1 kg / m^3 (line CD) and read velocity pressure 127 mmwg (line B)
Source:	base data from ABL Engineering Memoranda nomogram constructed

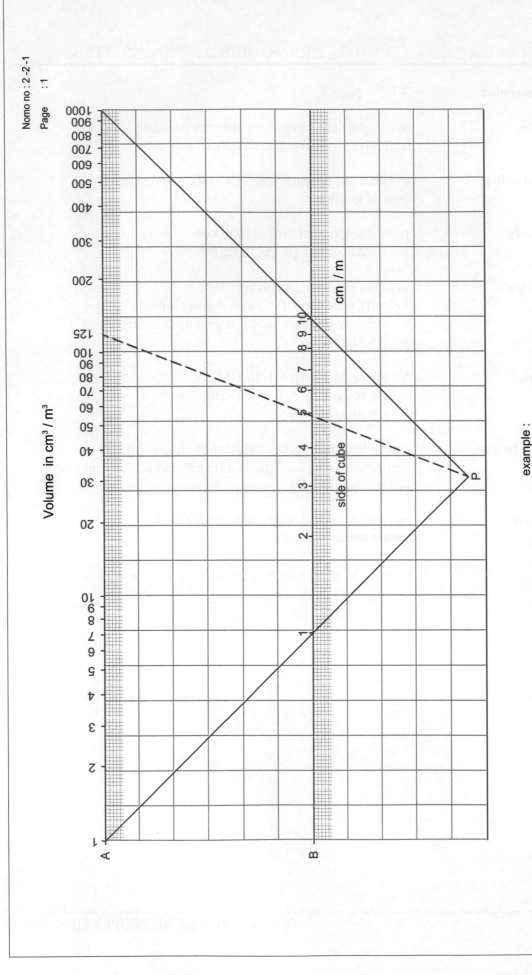

Volume in cm³ / m³

cm / m

side of cube

P

A

B

example :

side of cube = 5 m (line B)

volume of cube = 125 m³ (line A)

volume of sphere 5 m dia. = 0.524 x 125 = 65.5 m³

Scale : A - logartnmic scale for volume : 7 cm = 1-10, 10-100, 100-100 cm³ / m³

        B - constructed scale for side / dia of cube, sphere in cm / m

            for reading volume of sphere, multiply value on line A by 0.524

        P - point of reference

Nomogram for working out volumes of cubes & spheres

DEOLALKAR CONSULTANTS

Nomo no.:	2-2-1 page 2
Title:	Nomogram for working out volumes of Cubes and Spheres
Useful for:	finding out quickly volume of a cube or a sphere knowing its side or diameter
Inputs:	1 side 'a' of a cube in cm/m 2 diameter 'd' of a sphere in cm/m
Outputs:	volume of cube / sphere in cm^3/m^3 volume of a sphere = 0.524 × volume of cube where a = d
Scale:	A : Logarithmic scale for volume : 7 cms = 1-10, 10-100, 100-1000 B Specially constructed scale for side 'a' or diameter 'd' in cm/m
How to use:	From point of reference P, draw a line through side 'a' on line B to meet line A and read volume of cube on it. To arrive at volume of sphere of diameter d = a, multiply by 0.524

Example:

volume of cube side 5 m = 125 m^3

volume of sphere, diameter 5 m = 65.5 m^3

Source:	constructed

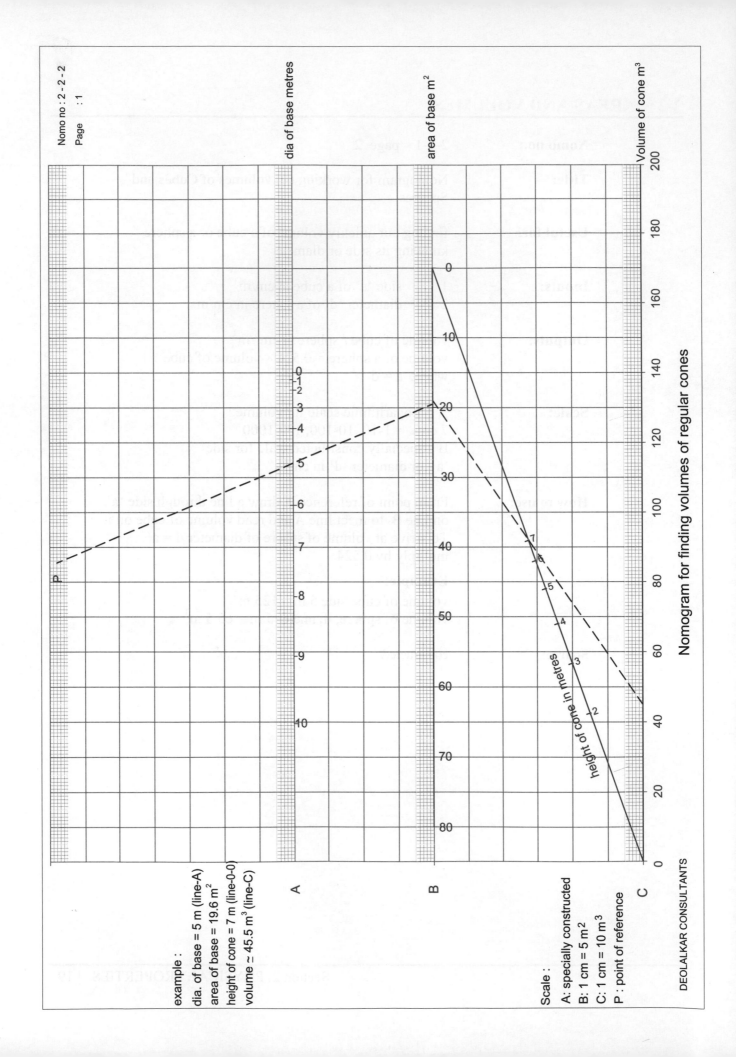

Nomo no : 2 - 2 - 2
Page : 1

dia of base metres

area of base m²

Volume of cone m³

Nomogram for finding volumes of regular cones

example :
dia. of base = 5 m (line-A)
area of base = 19.6 m²
height of cone = 7 m (line-0-0)
volume ≈ 45.5 m³ (line-C)

A

B

height of cone in metres

C

Scale :
A: specially constructed
B: 1 cm = 5 m²
C: 1 cm = 10 m³
P : point of reference

DEOLALKAR CONSULTANTS

Nomo no.:	2-2-2 page 2
Title:	Nomogram for finding volumes of regular cones
Useful for:	finding out volumes of circular and linear stockpiles for estimating their capacities
Inputs:	1 diameter of base circle in meters 2 height of cone in meters
Output:	volume of cone in m^3
Scale:	A : constructed scale for diameter of base in m B : 1 cm = 5 m^2 area of base C : 1 cm = 10 m^3 volume of cone
How to use:	From point of reference P, draw a line through given dia on line A, to meet line B; read area of base circle. From this point draw a line through given height of Cone to meet line C, and read volume of cone

Example:

dia of cone = 5 m (line A)
area of base = ~ 19.6 m^2 (line B)
height = 7 m (line 0-0)
volume = ~ 45.5 m^3

Source:	constructed

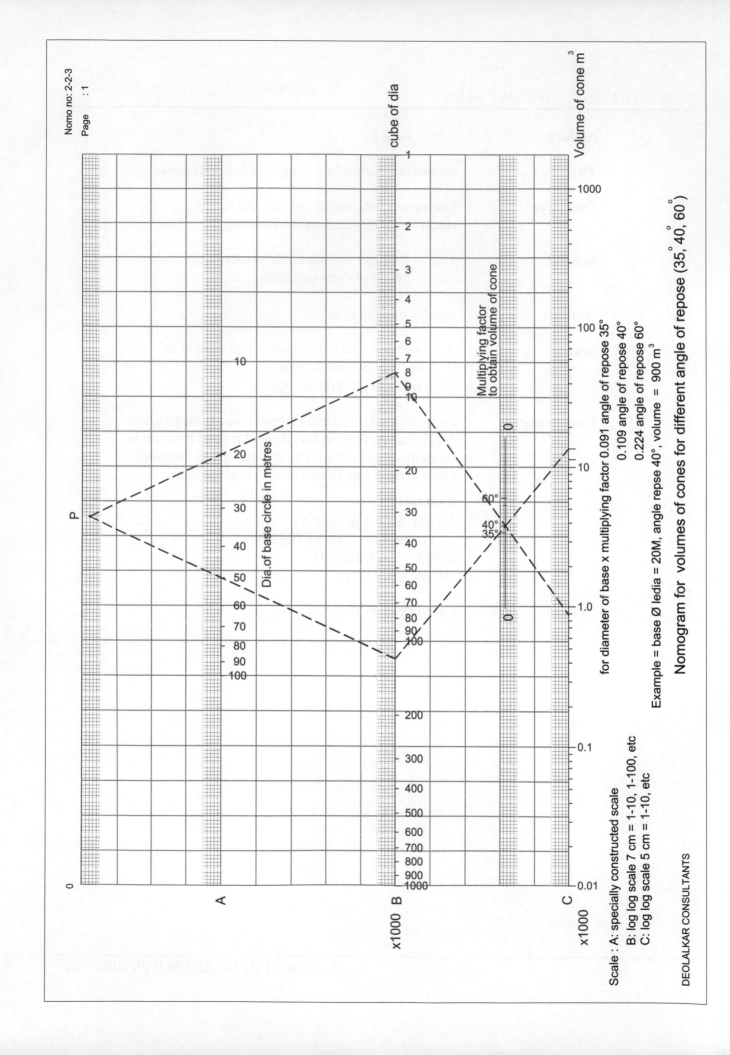

cube of dia

Volume of cone m³

1000

Multiplying factor
to obtain volume of cone

Dia. of base circle in metres

P

60°
40°
35°

for diameter of base × multiplying factor 0.091 angle of repose 35°
0.109 angle of repose 40°
0.224 angle of repose 60°

Example = base Ø ledia = 20M, angle repse 40°, volume = 900 m³

Nomogram for volumes of cones for different angle of repose (35°, 40°, 60°)

Scale : A: specially constructed scale
B: log log scale 7 cm = 1-10, 1-100, etc
C: log log scale 5 cm = 1-10, etc

A

B x1000

C x1000

DEOLALKAR CONSULTANTS

Nomo no.:	2-2-3 page 2	
Title:	Nomogram for working out volumes of cones with different angles of repose (35°, 40° and 60°)	
Useful for:	finding out volumes and hence capacities of circular and linear stock piles of materials like limestone, clinker etc.	
Inputs:	1 diameter of base circle in m 2 angle of repose of material	
Output:	Volume of cone in m^3	
Scale:	A : constructed scale for diameters 10-100 m B : logarithmic scale : 7 cm = 1-10, 10-100 etc. C : logarithmic scale : 4 cms = 0.01-0.1, 0.1-1 etc.	

How to use:

volume of cone $= 0.785 \times b^2 \times 1/3 \times h$

where b = dia. of base and h = height

h = 0.35 b for angle of repose 35°

    0.42 b -------- do --------- 40°

    0.87 b -------- do --------- 60°

therefore volume of cone $= 0.091 \times b^3$ for 35°

                      $= 0.109 \times b^3$ for 40°

                      $= 0.224 \times b^3$ for 60°

from point P draw a line through dia of base on line A to meet line B. From this point draw a line through appropriate multiplying factor on line 0-0 to meet Line C and read volume of cone in m^3

Example:
Dia. of cone = 20 m (line A)
Cube of dia. = 8000 m^3 (line B)
Angle of repose = 40° (line 0-0)
volume of cone = ~ 900 m^3 (line C)

Source: constructed

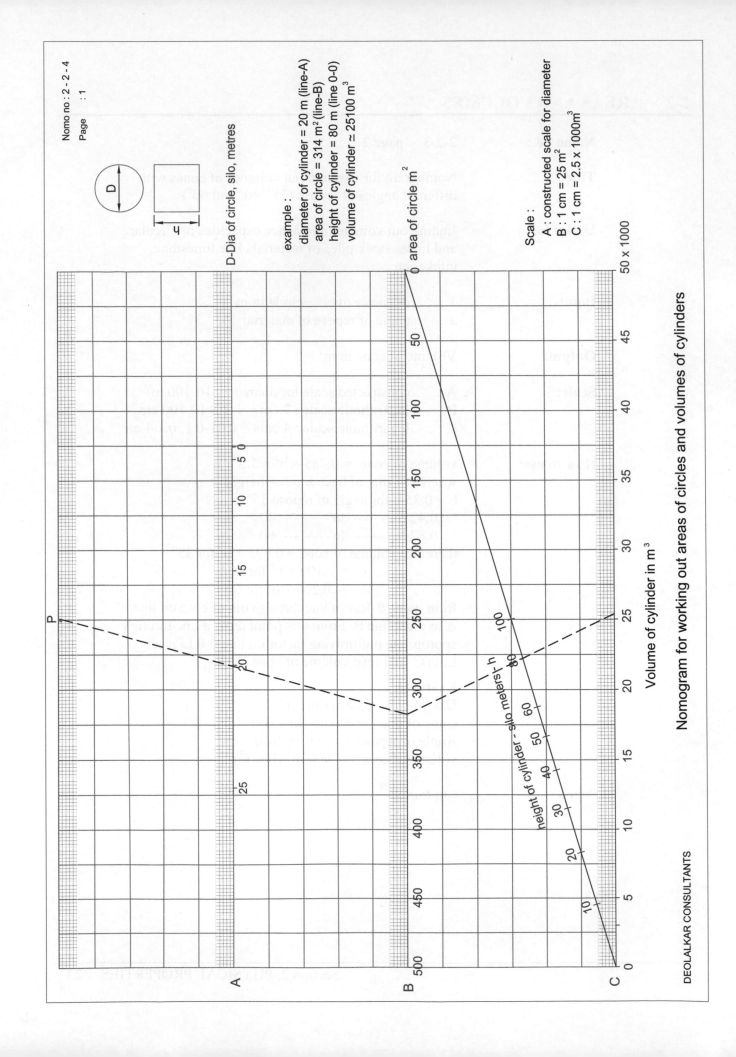

D-Dia of circle, silo, metres

example :
diameter of cylinder = 20 m (line-A)
area of circle = 314 m² (line-B)
height of cylinder = 80 m (line 0-0)
volume of cylinder ≃ 25100 m³

Scale :
A : constructed scale for diameter
B : 1 cm = 25 m²
C : 1 cm = 2.5 x 1000m³

area of circle m²

Volume of cylinder in m³

height of cylinder - silo meters - h

Nomogram for working out areas of circles and volumes of cylinders

DEOLALKAR CONSULTANTS

Nomo no.:	2-2-4 page 2
Title:	Nomogram for working out areas of circles and volumes of cylinders
Useful for:	volumes and capacities of storage silos
Inputs:	1 diameters of circles (silos) in m
	2 heights of cylinders (silos) in m
Output:	volumes of cylinders in m^3
Scale:	A : constructed scale for diameters of circles
	B : 1 cm = 25 m^2
	C : 1 cm = 2500 m^3
	P : Point of reference
How to use:	From point P draw a line through dia. of circle on line A to meet line B in say point D. From this point draw a line through height of cylinder on line 0-0 to meet line C at point E. Read volume of cylinder (silo)

Example:
dia. of cylinder = 20 m (line A)
area of circle = 314 m^2
height of cylinder = 80 m (line0-0)
volume of cylinder = ~ 25100 m^3

Source:	constructed

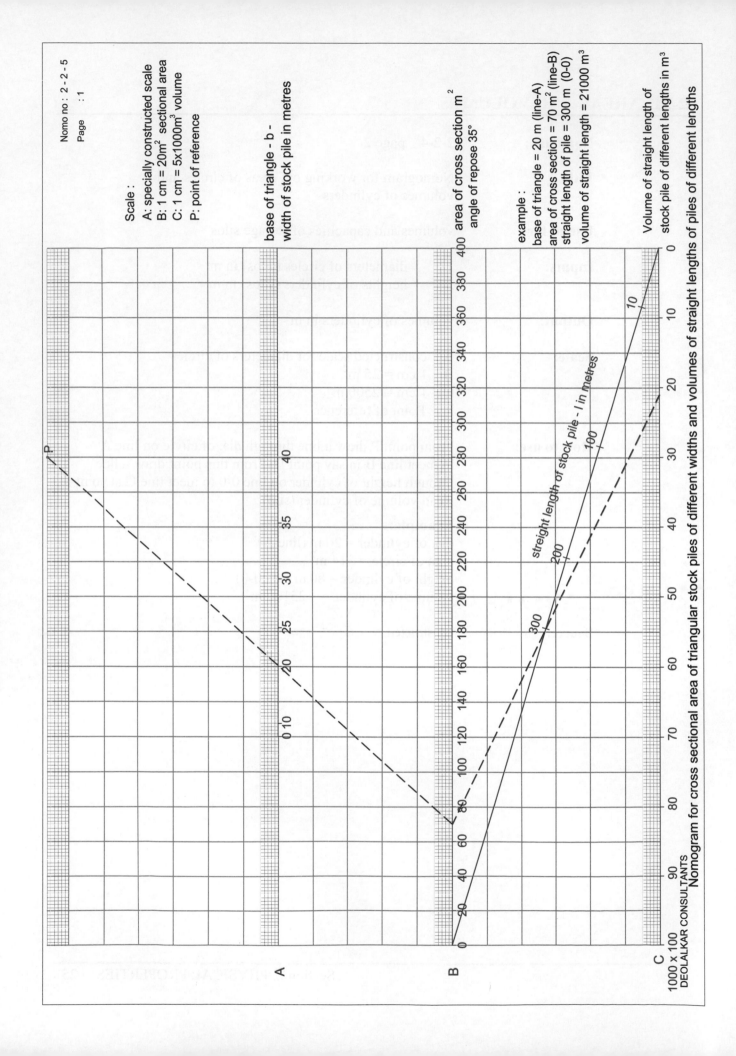

Nomo no : 2 - 2 - 5
Page   : 1

Scale :
A: specially constructed scale
B: 1 cm = 20m² sectional area
C: 1 cm = 5x1000m³ volume
P: point of reference

base of triangle - b -
width of stock pile in metres

area of cross section m²
angle of repose 35°

example :
base of triangle = 20 m (line-A)
area of cross section = 70 m² (line-B)
straight length of pile = 300 m (0-0)
volume of straight length = 21000 m³

Volume of straight length of
stock pile of different lengths in m³

straight length of stock pile - l in metres

Nomogram for cross sectional area of triangular stock piles of different widths and volumes of straight lengths of piles of different lengths

DEOLALKAR CONSULTANTS

Nomo no.:	2-2-5 page 2
Title:	Nomogram for working out cross sectional areas of stock piles of different widths and volumes of straight lengths of piles of different lengths
Useful for:	working out capacities of linear stock piles of crushed limestone, clinker, coal etc used in stacker reclaimer systems
Inputs:	1 width of pile at base in m ' b' 2 angle of repose 35 $^{\circ}$ 3 height of pile = 0.35 × 'b' 4 cross section of pile = 0.175 ×' b'^{2} 5 length of pile – straight portion 'l' m
Output:	volume of straight length of stock pile in m^3 Volume = 0.175 × b^2 × l
Scale:	P : point of reference A : constructed scale for width of pile 'b' in m B : 1 cm = 20 m^2 area of cross section C : 1 cm = 5000 m^3 volume of straight length
How to use:	from point of reference P draw a line through width of pile on line A and extend to line B and read cross sectional area (point D). From this point draw a line through length of pile (straight length) on line 0-0 to meet line C at point E and read volume in m^3 **Example:** width of pile = 20 m (line A) cross sectional area = 70 m^2 (line B) length of pile = 300 m (line 0-0) volume of pile = 21000 m^3 (line C) – straight length only
Source:	constructed

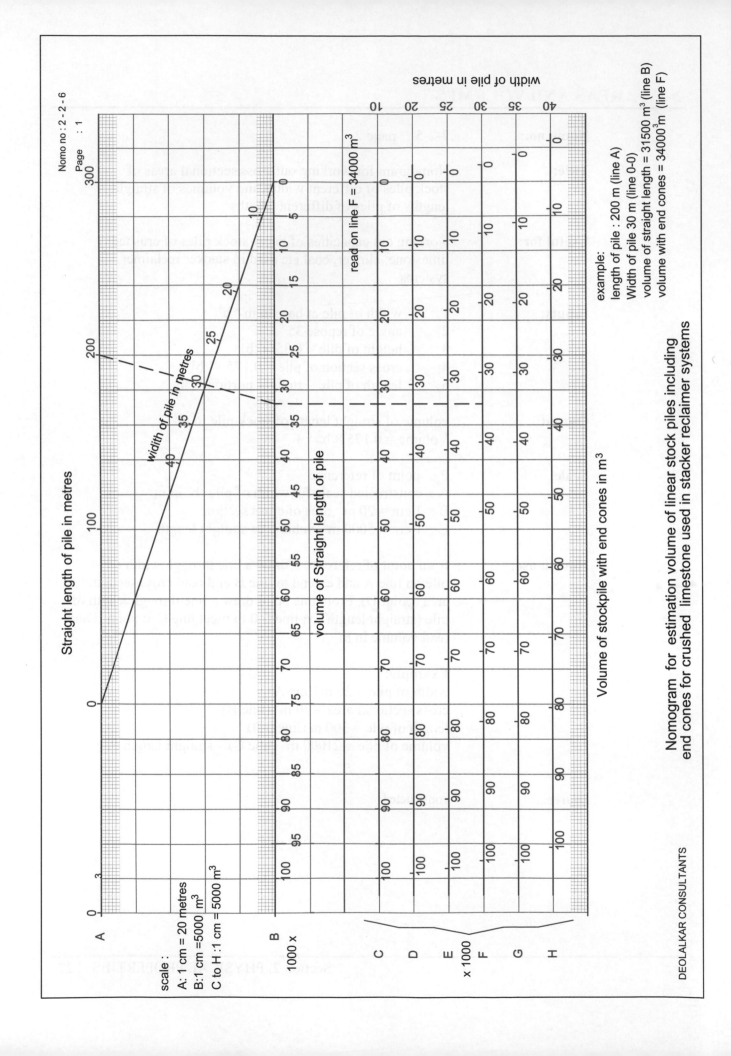

Nomo no : 2 - 2 - 6
Page   : 1

Straight length of pile in metres

scale :
A: 1 cm = 20 metres
B:1 cm =5000 m³
C to H :1 cm = 5000 m³

width of pile in metres

width of pile in metres

Width of pile in metres

read on line F = 34000 m³

volume of Straight length of pile

Volume of stockpile with end cones in m³

example:
length of pile : 200 m (line A)
Width of pile 30 m (line 0-0)
volume of straight length = 31500 m³ (line B)
volume with end cones = 34000³ m (line F)

Nomogram for estimation volume of linear stock piles including
end cones for crushed limestone used in stacker reclaimer systems

DEOLALKAR CONSULTANTS

Nomo no.:	2-2-6 page 2
Title:	Nomogram for estimating volumes of linear stock piles including cones, for crushed limestone, used in stacker reclaimer systems
Useful for:	estimating capacities of stockpiles and conversely dimensions of stock piles for planning layouts
Inputs:	1 width 'b' of stock pile in metres 2 length of straight (triangular) portion 'l' of stock pile in meters 3 assumed angle of repose 35 $^\circ$
Output:	1st stage: volume of straight length 2nd stage: volume of pile with end cones

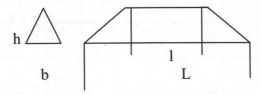

Scale:	A : 1 cm = 20 m B : 1 cm = 5000 m^3 C to H = 1 cm = 5000 m^3
How to use:	Draw a line from length of pile 'l' on line A thru width of pile 'b' on line 0-0 and extend to meet line B and read volume of straight length without cones for volume with end cones draw a perpendicular from this point to appropriate line showing width of pile and read total volume with end cones in m^3 **Example:** st. length of pile 'l' = 200 m (line A) width of pile = 30 m (line 0-0) volume of st. length = ~ 31500 m^3 (line B) volume of pile with end cones = ~ 34000 m^3 (line F)
Source:	constructed

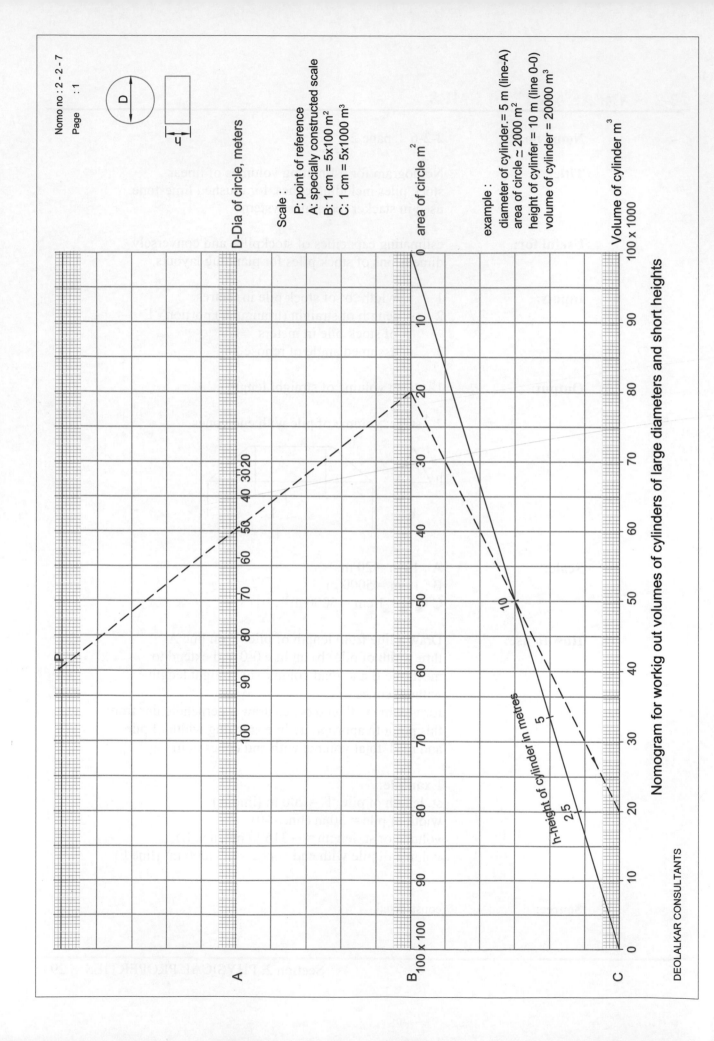

Nomo no : 2 - 2 - 7
Page   : 1

D-Dia of circle, meters

Scale :
P: point of reference
A: specially constructed scale
B: 1 cm = 5x100 m²
C: 1 cm = 5x1000 m³

area of circle m²

example :
diameter of cylinder.= 5 m (line-A)
area of circle ≃ 2000 m²
height of cylinfer = 10 m (line 0-0)
volume of cylinder = 20000 m³

Volume of cylinder m³

Nomogram for workig out volumes of cylinders of large diameters and short heights

DEOLALKAR CONSULTANTS

Nomo no.:	2-2-7 page 2
Title:	Nomogram for working out volumes of cylinders of large diameters and short heights
Useful for:	working out capacities of clinker stock piles
Inputs:	1 diameter of base circle in m 2 height of cylinder in m
Output:	Volume of cylinder in m^3
Scale:	P : point of reference A : specially constructed scale for dia. of base B : 1 cm = 500 m^2 C : 1 cm = 5000 m^3
How to use:	Draw a line from point of reference P through diameter of base on line A, to meet line B at point D. From this point draw a line through height of cylinder on line 0-0 to meet line C at point E. Read volume of cylinder in m^3

Example:
dia. of cylinder = 50 m (line A)
area of circle = ~2000 m^2
height of cylinder = 10 m (line 0-0)
volume of cylinder = $\simeq$ 20000 m^3

Source:	constructed

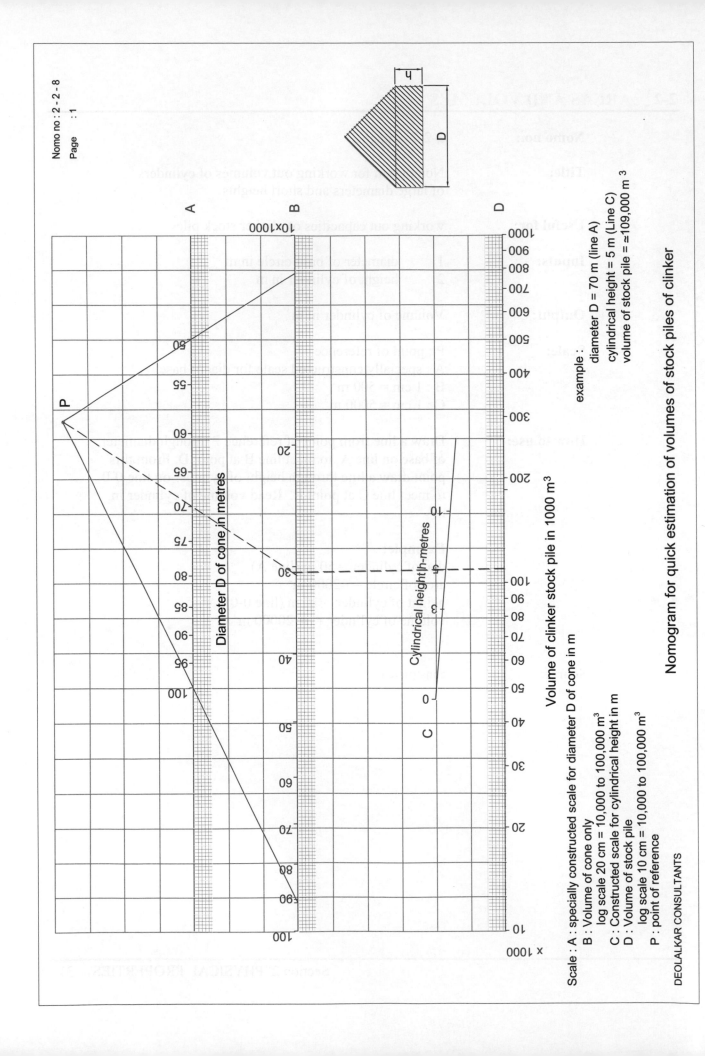

Nomo no : 2 - 2 - 8
Page   : 1

example :

diameter D = 70 m (line A)
cylindrical height = 5 m (Line C)
volume of stock pile = ≃109,000 m³

Volume of clinker stock pile in 1000 m³

Nomogram for quick estimation of volumes of stock piles of clinker

Scale : A : specially constructed scale for diameter D of cone in m
        B : Volume of cone only
            log scale 20 cm = 10,000 to 100,000 m³
        C : Constructed scale for cylindrical height in m
        D : Volume of stock pile
            log scale 10 cm = 10,000 to 100,000 m³
        P : point of reference

DEOLALKAR CONSULTANTS

Nomo no.:	2-2-8 page 2
Title:	Nomogram for quick estimation of volumes of stock piles of clinker
Useful for:	Designing clinker stockpiles
Inputs:	1 dia. of stock pile 50-100 m 2 cylindrical height 3-10 m 3 angle of repose assumed 35 $^\circ$
Output:	Total volume including cylindrical height in m^3
Scale :	A : specially constructed scale for dia. D of cone in m B : log scale : 20 cm = 10000 – 100,000 m^3 vol. of cone only C : specially constructed scale for height of cylindrical portion in m D : log scale : 10 cm = 10000 – 100,000 m^3 vol. of total pile P : point of reference
How to use:	Draw a line from P thru base circle dia (line A) to meet line B read volume of cone only (line B) from this point draw line thru height of cylindrical portion (line C) and extend to meet line D and read volume of total pile. **Example:** dia. at base = 70 m (line A) cylindrical height = 5 m (line C) volume of total pile = $\simeq$ 109000 m^3 (line D)
Source:	constructed

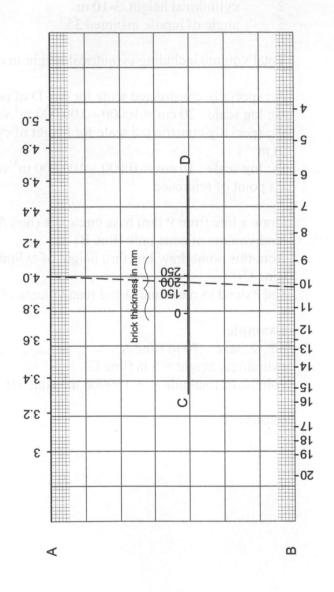

Diameter inside shell in m

Clear cross sectional area m²

brick thickness in mm

example : kiln dia = 4 m (line A)

brick thick = 200 mm (line CD)

clear area = ≈ 10.2² m (line B)

Scale : A, B & C : Specially constructed scales

Nomogram for working out clear cross sectional areas of brick lined rotary kilns and coolers

DEOLALKAR CONSULTANTS

Nomo no.:	2-2-9 page 2
Title:	Nomogram for working out clear cross sectional areas of brick lined rotary kilns and coolers
Useful for:	working out thermal load, clear volume, sp. outputs etc.
Inputs:	1 nominal diameters of kiln, cooler etc. 2 thickness of brick lining in mm
Output:	Clear cross sectional area in m^2
Scale:	A : specially constructed scale for dia. in m B : specially constructed scale for area in m^2 C-D : points showing brick thickness in mm
How to use:	Draw a line from dia of kiln on line A thru brick thickness on line CD to meet line B and read clear cross sectional area **Example:** dia. of kiln = 4 m (line A) brick thickness = 200 mm (line CD) clear cross sectional area = ~ 10.2 m^2 (line B)
Source:	constructed

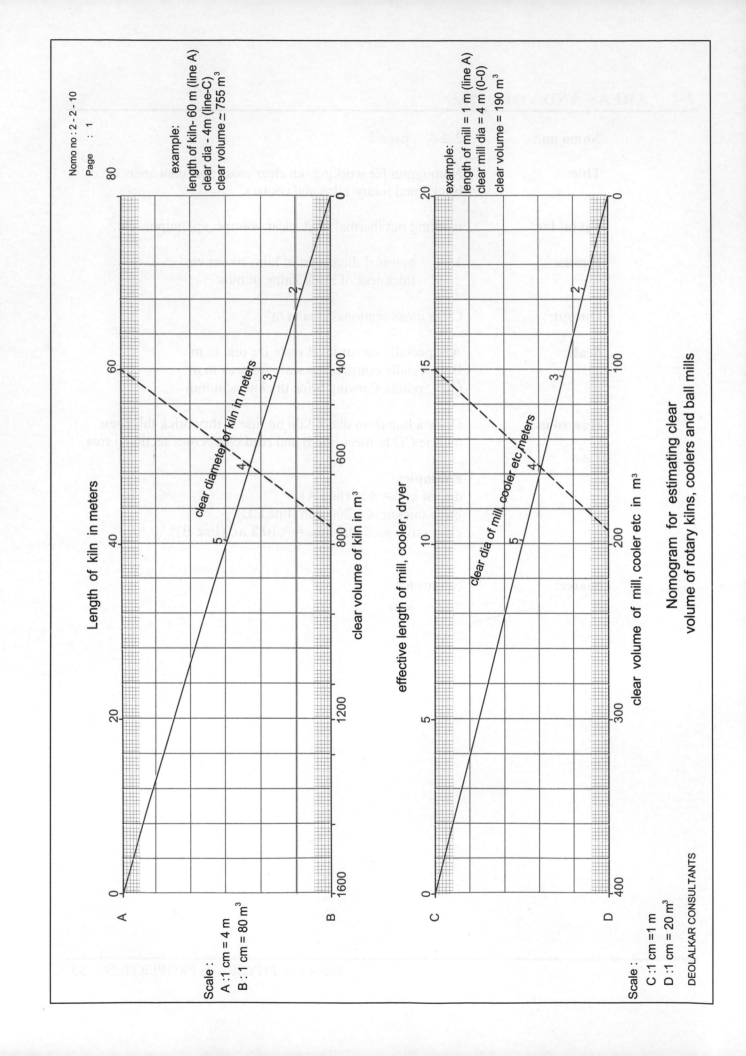

Nomo no : 2 - 2 - 10

Page : 1

example:

length of kiln- 60 m (line A)

clear dia - 4m (line-C)

clear volume ≈ 755 m³

Length of kiln in meters

80

60

40

20

0

A

clear diameter of kiln in meters

clear volume of kiln in m³

0   400   600   800   1200   1600

B

Scale :

A :1 cm = 4 m

B : 1 cm = 80 m³

example:

length of mill = 1 m (line A)

clear mill dia = 4 m (0-0)

clear volume = 190 m³

effective length of mill, cooler, dryer

20

15

10

5

0

C

Clear dia of mill, cooler etc meters

clear volume of mill, cooler etc in m³

0   100   200   300   400

D

Scale :

C :1 cm =1 m

D :1 cm = 20 m³

**Nomogram for estimating clear
volume of rotary kilns, coolers and ball mills**

DEOLALKAR CONSULTANTS

Nomo no.:	2-2-10 page 2
Title:	Nomogram for estimating clear volumes of rotary kilns coolers and ball mills etc.
Useful for:	quick estimation of grinding media charge, brick lining etc.
Inputs:	1 clear diameters inside of lining, lining plates etc. 2 effective length in metres
Output:	clear volume in m^3

Scale:	1 A : 1 cm = 4 m
	        B : 1 cm = 80 m^3
	2 C : 1 cm = 1 m
	        D : 1 cm = 20 m^3

How to use:	Draw a line from length of kiln on line A thru clear dia. on line 0-0 and extend to meet line B and read clear volume in m^3 **Example:** kiln length = 60 m (line A) clear dia = 4 m (line 0-0) clear volume = ~ 760 m^3 mill effective length = 15 m (line C) clear dia = 4 m (line 0'-0') effective volume = ~190 m^3
Source:	constructed

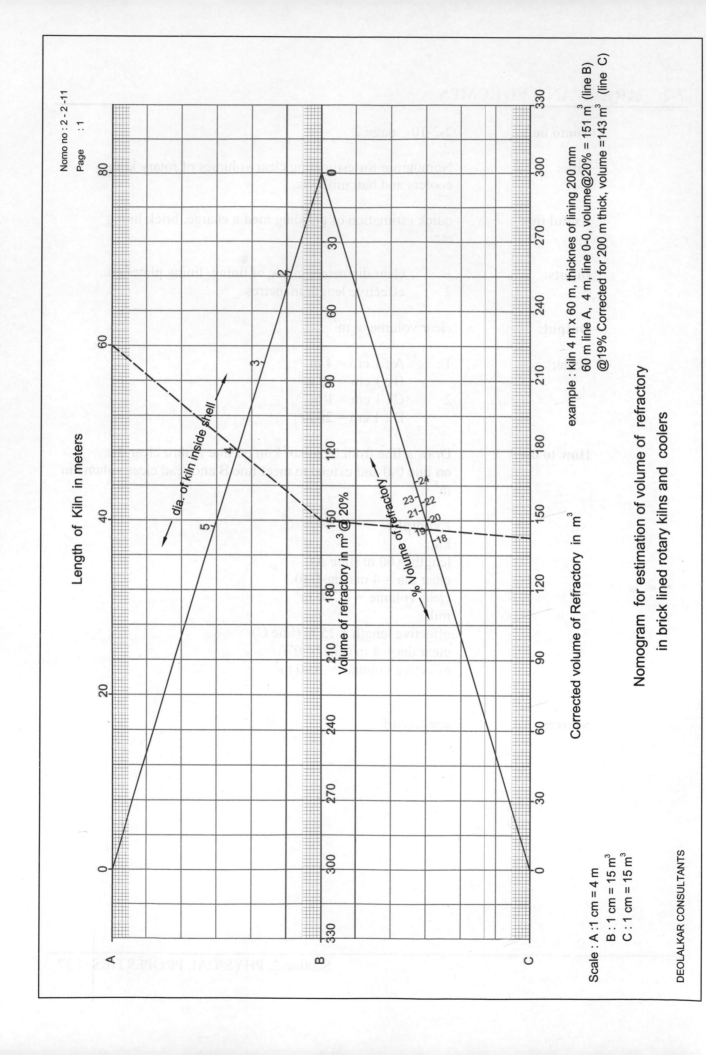

Nomo no : 2 - 2 -11
Page : 1

Length of Kiln in meters

dia. of kiln inside shell

Volume of refractory in m³ @ 20%

% Volume of refractory

Corrected volume of Refractory in m³

example : kiln 4 m x 60 m, thicknes of lining 200 mm
60 m line A, 4 m, line 0-0, volume@20% = 151 m³ (line B)
@19% Corrected for 200 m thick, volume =143 m³ (line C)

Nomogram for estimation of volume of refractory
in brick lined rotary kilns and coolers

Scale : A :1 cm = 4 m
B : 1 cm = 15 m³
C : 1 cm = 15 m³

DEOLALKAR CONSULTANTS

Nomo no.:	2-2-11 page 2
Title:	Nomogram for estimation of volume of refractory in brick lined rotary kilns and coolers
Useful for:	Quick estimation of quantity of refractory
Inputs:	1 diameter of kiln inside shell 2-5.2 m 2 length of kiln in m 3 volume calculated on basis of 20 % volume of kiln
Output:	Volume of refractory in m^3
Scale:	A : 1 cm = 4 m length of kiln metres B : 1 cm = 15 m^3 volume of refractory in m^3 C : 1 cm = 15 m^3 volume of refractory in m^3
How to use:	Given kiln 4 m dia × 60 m long draw a line from 60 on line A thru 4 m on line 0-0 to meet line B. Read refractory volume $\simeq$ 150 m^3 for a refractory lining of 200 mm thick, actual volume would be 19 %. From line B, draw a line thru 19 % on line 0-0' to meet line C and read corrected volume = ~ 143 m^3
Source:	constructed

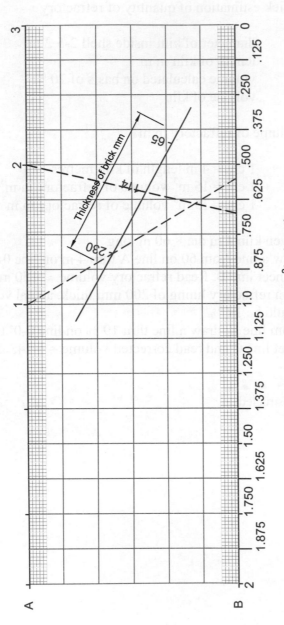

Diameter of duct in meters

Volume of brick lining in m³/m length

Thickness of brick mm

example :

dia of duct -2.0 m (line A)

thickness of lining -114 mm (0–0)

volume of lining - 0.675 m³/ m

Nomogram for estimation of refractory
lining in ducts in m³/m length

scale :

A : 1 cm = 0.25 m.

B : 1 cm = 0.125 m³

DEOLALKAR CONSULTANTS

Nomo no.:	2-2-12 page 2
Title:	Nomogram for estimation of refractory lining in ducts in m^3/m length
Useful for:	Estimating quantities of refractory lining in tertiary ducts, calciners etc.
Inputs:	1 diameter of duct in m 2 brick/lining thickness in mm
Output:	volume of brick lining in m^3/m length of duct
Scale:	A : 1 cm = 0.25 m dia B : 1 cm = 0.125 m^3
How to use:	draw a line from given dia of duct on line A thru thickness of lining on line 0-0 and extend to meet line B and read volume of lining in m^3/m

Example:
dia. of duct = 2.0 m inside shell (line A)
thickness of refractory = 114 mm (line 0-0)
volume of refractory = 0.675 m^3 per metre

Source:	constructed

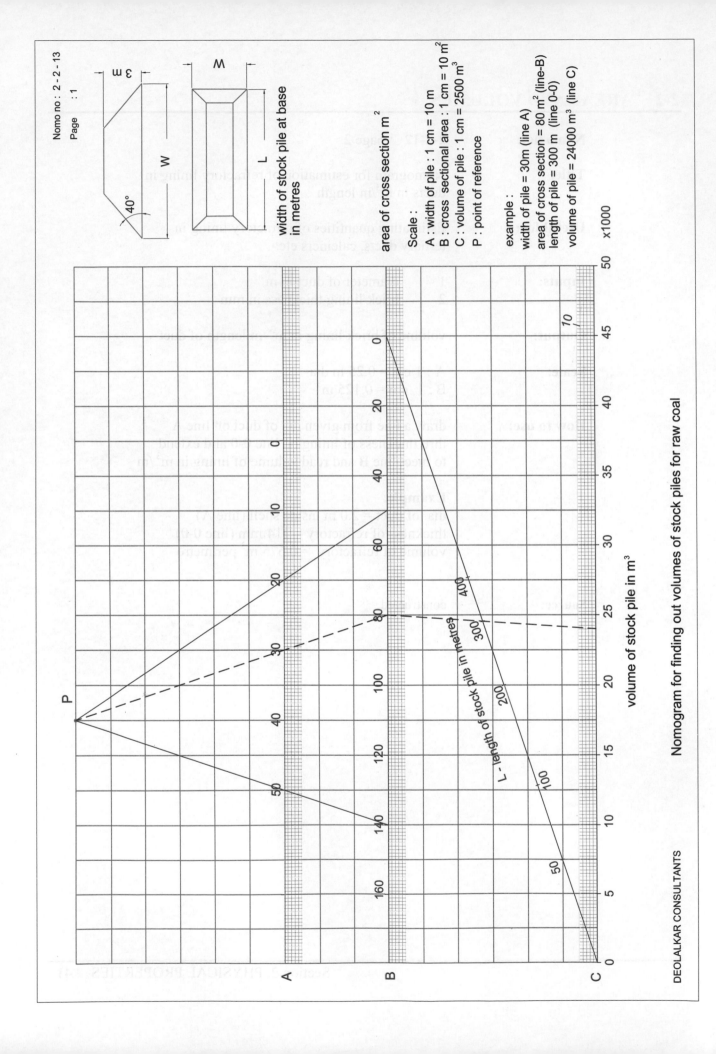

Nomo no : 2 - 2 - 13
Page : 1

width of stock pile at base in metres

area of cross section m²

Scale :

A : width of pile : 1 cm = 10 m
B : cross sectional area : 1 cm = 10 m²
C : volume of pile : 1 cm = 2500 m³
P : point of reference

example :

width of pile = 30m (line A)
area of cross section = 80 m² (line-B)
length of pile = 300 m (line 0-0)
volume of pile = 24000 m³ (line C)

volume of stock pile in m³

L - length of stock pile in metres

Nomogram for finding out volumes of stock piles for raw coal

DEOLALKAR CONSULTANTS

Nomo no.:	2-2-13 page 2
Title:	Nomogram for finding out volumes of linear Stockpiles for raw coal
Useful for:	Estimating requirements of space to be provided for designing storage and handling systems for raw coal.
Inputs:	Basis:
	a Cross section of stock pile – trapezium in shape
	b height of stockpile 3 metres (restricted to prevent possibility of coal catching fire in dry months)
	c angle of repose of wet coal 40°
	1 width of stock pile at base - W- in metres
	2 length of stock pile at base –L –in metres
Output:	Volume of stock pile ignoring reduction in volume at ends due slopes as it is small.
Scale:	A : width –W-: 1 cm = 10m
	B : cross sectional area : 1 cm = 10 m^2
	C : volume of stock pile : 1 cm = 2500 m^3
	P : point of reference
How to use;	From point of reference P draw a line thru the width of pile on Line A and extend to meet line B to read cross section. From This point draw a line to the length of the pile on line 0-0 and Extend to meet line C and read volume in m^3
	Example:
	width of pile : 30 m (line A)
	length of pile : 300 m (line 0-0)
	volume of stock pile : = ~ 24000 m^3 (line C)
Source:	constructed

Nomogram for finding sin Θ/2, given Θ/2 half angle at centre

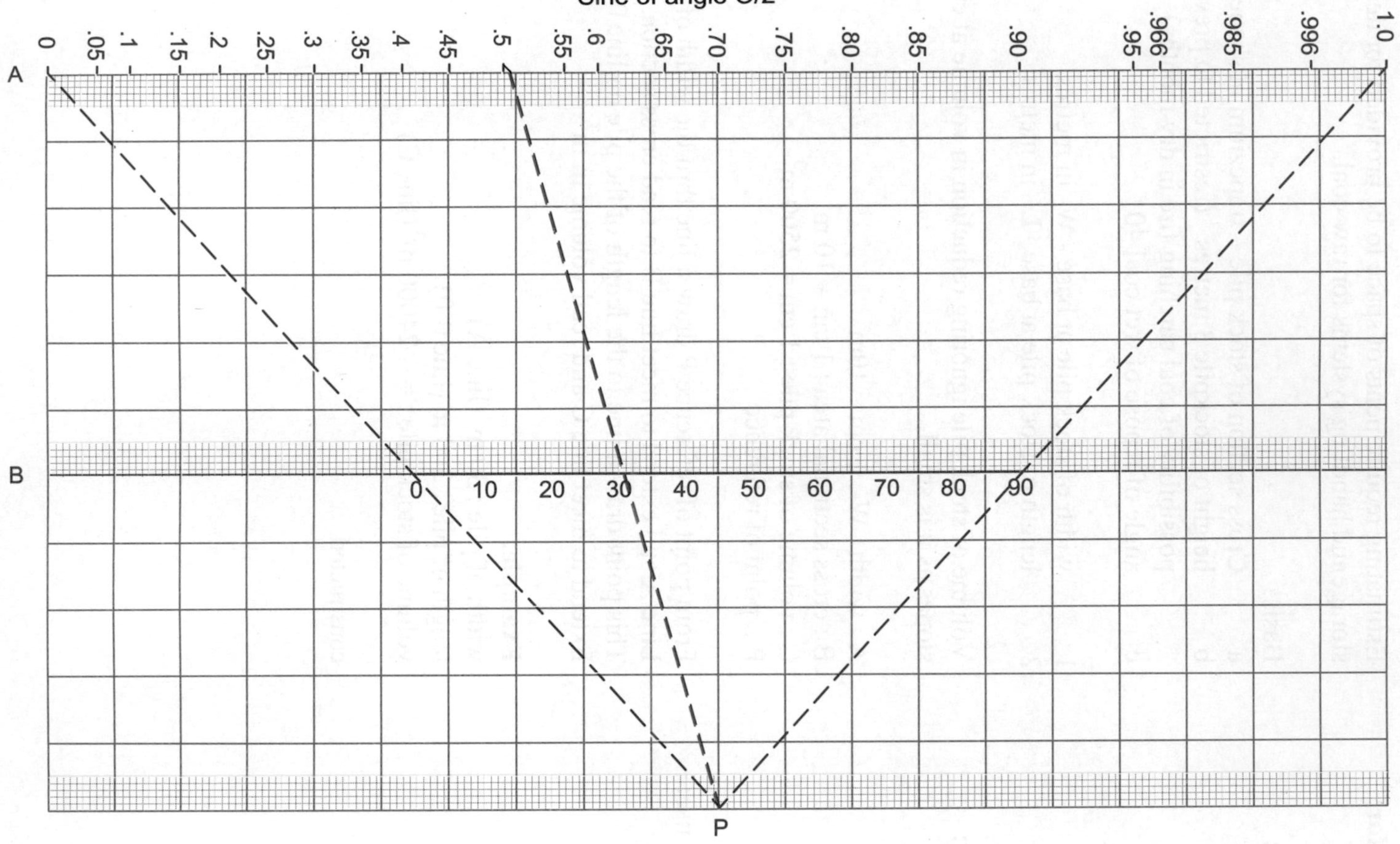

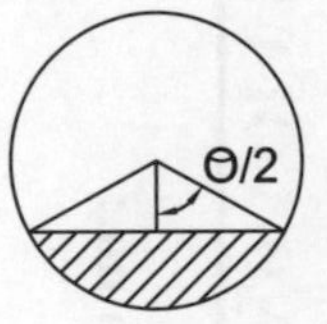

Scale : A : Specially constructed for sin Θ/2
        B : 1 cm = 10° Θ/2

Sine of angle Θ/2

half angle Θ/2 at centre

example : Θ/2 = 30° (line B)
          sin Θ/2 = 0.5 (line A)

DEOLALKAR CONSULTANTS

Nomo no.:	2-3-1 page 2
Title:	Nomogram for finding sin θ/2, given θ/2 half angle at center
Useful for:	finding % loading and related calculations for ball mills and rotary kilns and coolers
Inputs:	half angle θ/2 at center
Output:	sine of half angle
Scale:	A : specially constructed B : 1 cm = 10 $^\circ$ P point of reference
How to use:	From reference point P draw a line thru half angle on line B and extend to line A and read sine of half angle

Example:
half angle θ/2 at center = 30 $^\circ$ (line B)
sine θ/2 = 0.5 (line A)

Source:	constructed

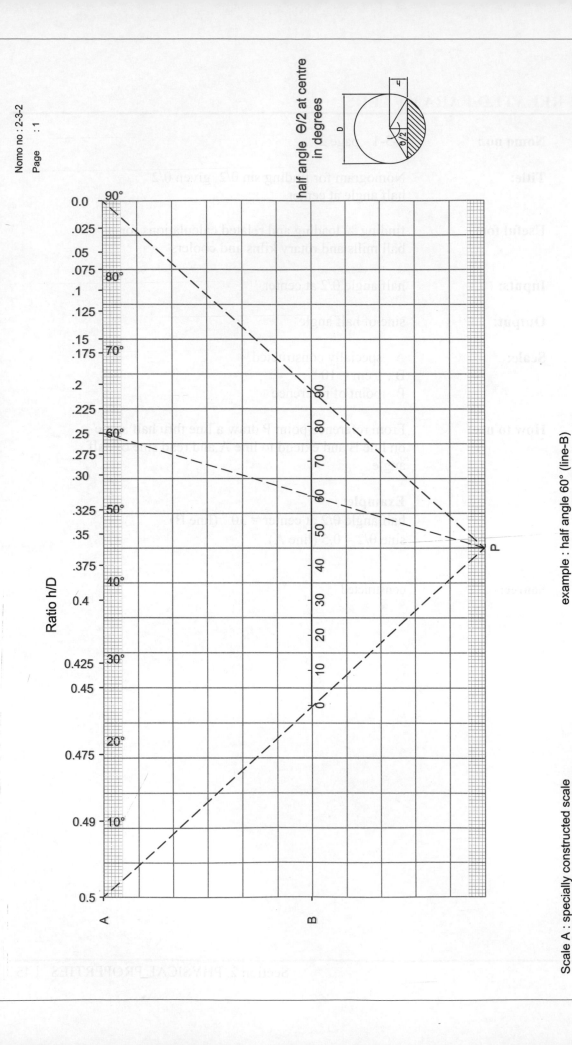

Ratio h/D

half angle Θ/2 at centre
in degrees

Scale A : specially constructed scale

B : 1 cm = 10° half angle at centre

P : point of reference

example : half angle 60° (line-B)
ratio h/D = 0.25 (line-A)

Nomogram for finding out ratio h/D for given of half angle Θ/2 at centre

Nomo no.:	2-3-2 page 2
Title:	Nomogram for finding out ratio h/D for given θ/2 at centre
Useful for:	finding % loading and other parameters related to power of ball mills, and rotary kilns and coolers etc.
Inputs:	half angle at center θ/2 in degrees
Output:	ratio h/D h is empty height from centre of mill / kiln D is dia of mill/kiln inside of liners, bricks etc. (see sketch)
Scale:	A : specially constructed scale for h/D B : 1 cm = 10 ° half angle P : point of reference
How to use:	draw a line from point of reference P thru given half angle on line B and extend to meet line A and read ratio h/D

Example:
half angle at centre θ/2 = 60° (line B)
ratio h/D = 0.25 (line A)

Source:	constructed

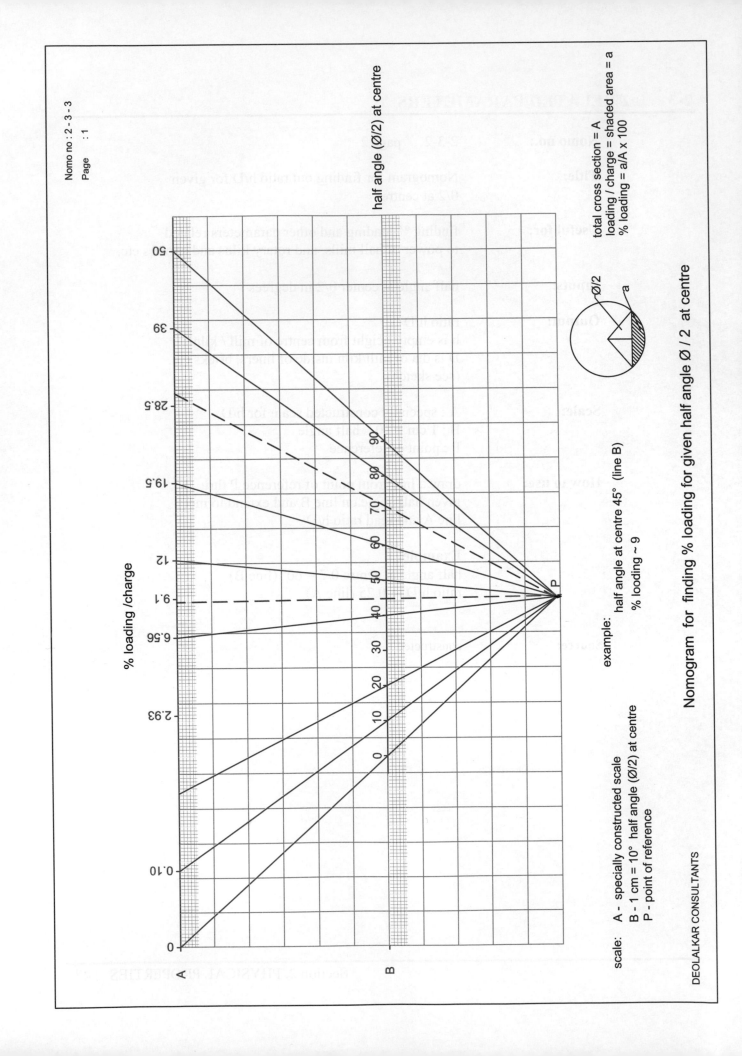

half angle (Ø/2) at centre

% loading /charge

total cross section = A
loading / charge = shaded area = a
% loading = a/A x 100

P

example: half angle at centre 45° (line B)
           % loading ~ 9

scale: A - specially constructed scale
         B - 1 cm = 10° half angle (Ø/2) at centre
         P - point of reference

Nomogram for finding % loading for given half angle Ø / 2 at centre

DEOLALKAR CONSULTANTS

Nomo no.:	2-3-3 page 2
Title:	Nomogram for finding % loading for given half angle θ/2 at centre
Useful for:	finding charge inside mills, kilns etc.
Inputs:	half angle θ/2 at cntre
Output:	% loading See sketch
Scale:	A : specially constructed scale for % loading B : 1 cm = 10° half angle θ/2 at centre P : point of reference
How to use:	draw a line from point of reference, P thru given half angle on line B to met line A and read % loading or charge

Example:
% loading = 30   (line A)
half angle θ/2 = 72 °

half angle θ/2 = 45 ° (line B)
% loading = ~ 9 % (line A)

Source:	constructed

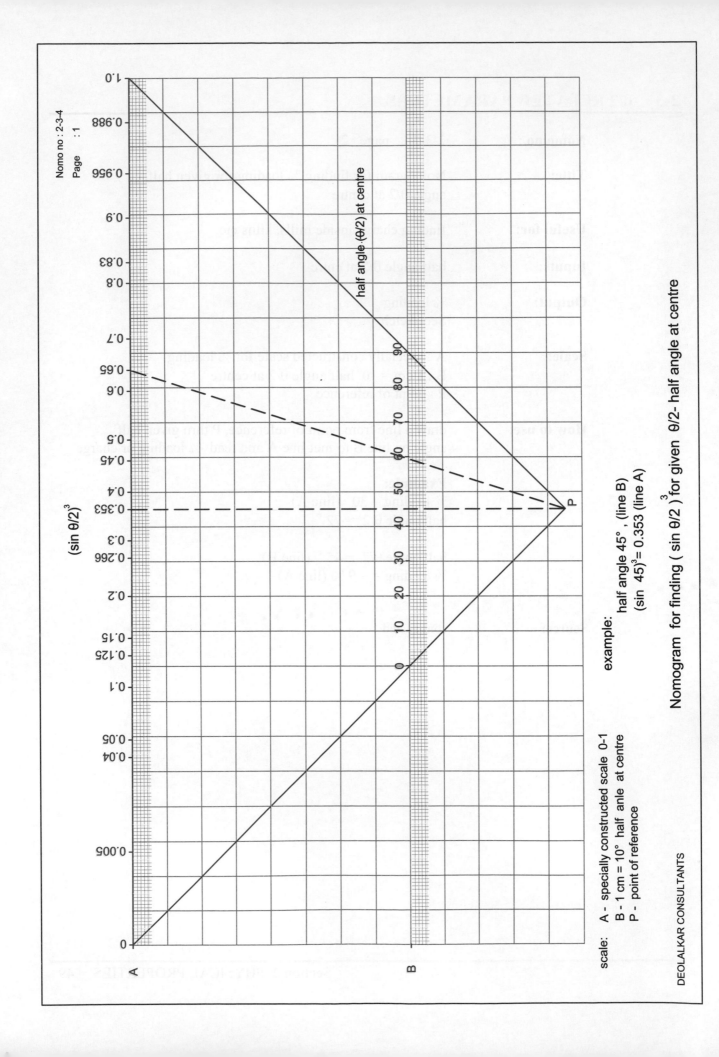

Nomo no : 2-3-4
Page   : 1

$(\sin \theta/2)^3$

half angle ·(θ/2) at centre

scale: A - specially constructed scale 0-1
         B - 1 cm = 10° half anle at centre
         P - point of reference

example: half angle 45° , (line B)
            $(\sin 45)^3$ = 0.353 (line A)

Nomogram for finding (sin θ/2)³ for given θ/2- half angle at centre

DEOLALKAR CONSULTANTS

Nomo no.:	2-3-4 page 2
Title:	Nomogram for finding $(\sin \theta/2)^3$ for given half angle θ/2 at centre
Useful for:	finding load power for rotary kilns, coolers etc.
Input:	θ/2 half angle at center
Output:	$(\sin \theta/2)^3$
Scale:	A : specially constructed scale for 0-1 B : 1 cm = 10 ° P : point of reference
How to use:	draw a line from reference point P thru given half angle on line B and extend to meet line A and read $(\sin \theta/2)^3$

Example:
half angle θ/2 at centre = 45 ° (line B)
$(\sin \theta/2)^3$ = 0.353 (line A)

Source:	constructed

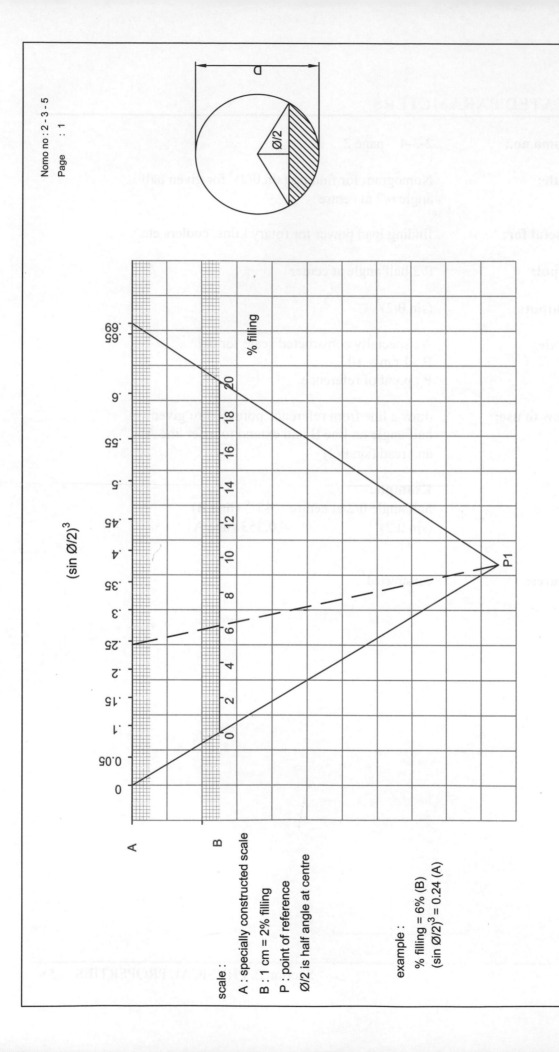

D

$\emptyset/2$

$(\sin \emptyset/2)^3$

% filling

.69
.65
.6
.55
.5
.45
.4
.35
.3
.25
.2
.15
.1
0.05
0

A

B

P1

scale :

A : specially constructed scale

B : 1 cm = 2% filling

P : point of reference

$\emptyset/2$ is half angle at centre

example :

% filling = 6% (B)

$(\sin \emptyset/2)^3 = 0.24$ (A)

Nomogram for finding $(\sin \emptyset/2)^3$ given % filling

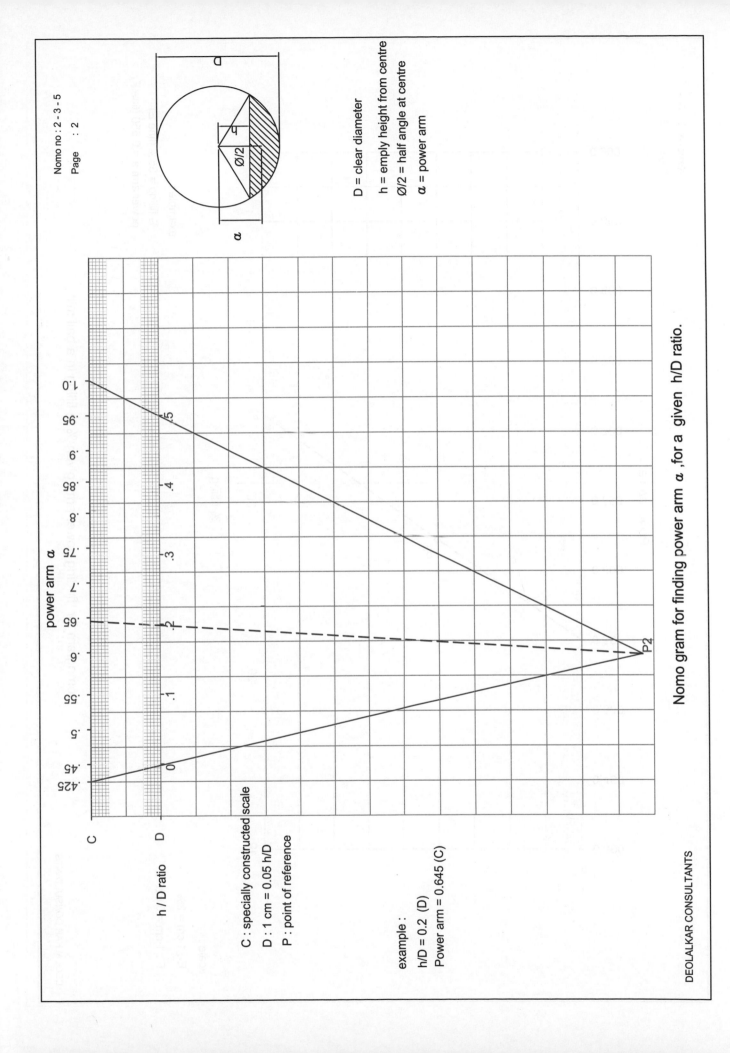

Nomo no : 2 - 3 - 5
Page : 2

D = clear diameter

h = empty height from centre

Ø/2 = half angle at centre

α = power arm

power arm α

h / D ratio

C : specially constructed scale

D : 1 cm = 0.05 h/D

P : point of reference

example :

h/D = 0.2 (D)

Power arm = 0.645 (C)

Nomo gram for finding power arm α ,for a given h/D ratio.

DEOLALKAR CONSULTANTS

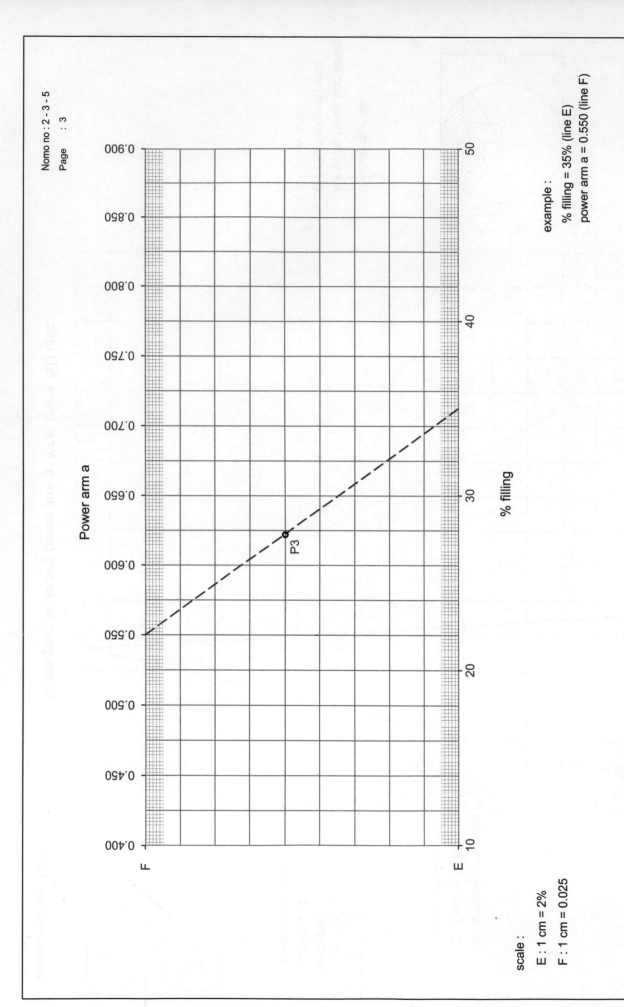

Power arm a

0.900 0.850 0.800 0.750 0.700 0.650 0.600 0.550 0.500 0.450 0.400

F

50 40 30 20 10

E

P3

% filling

example :

% filling = 35% (line E)
power arm a = 0.550 (line F)

Nomogram for finding power arm, given % in filling in a ball mill

scale :

E : 1 cm = 2%

F : 1 cm = 0.025

DEOLALKAR CONSULTANTS

Nomo no.:	2-3-5 page 4
Title:	Nomogram for finding $(\text{Sin }\theta/2)^{3.5}$ and power arm given % filling and h/D ratio
Useful for:	working out power for rotary kiln, rotary cooler and for power for ball mills
Inputs:	1 % filling 2 h/D ratio
Outputs:	1 $(\sin \theta/2)^3$ 2 power arm 3 h/D ratio

Scale:	page 1 A : specially constructed scale for sin $(\theta/2)^3$
	           B : 1 cm = 2 % filling
	           P1 : point of reference
	
	page 2 C : specially constructed scale for power arm
	           D : 1 cm = 0.05 h/D ratio
	           P2 : point of reference
	
	Page 3 E : 1 cm = 2 % filling
	           F : 1 cm = 0.025
	           P3 : point of reference

How to use:	1 draw a line from reference point P1 thru % filling on line B and extend to line A and read % filling 2 draw a line from reference point P2 thru h/D ratio on line D and extend to line C and read power arm 3 draw a line from given % filling on line E and extend thru point P3 to meet line F and read power arm **Example:** 1 % filling = 6 % (line B) $(\sin \theta/2)^3$ = ~ 0.24 (line A) 2 h/D ratio = 0.2 (line D) power arm = ~ 0.645 (line C) 3 filling = 35 % (line E) power arm = 0.55 (line F)
Source:	base data from FLS Manual for ball mills nomogram constructed

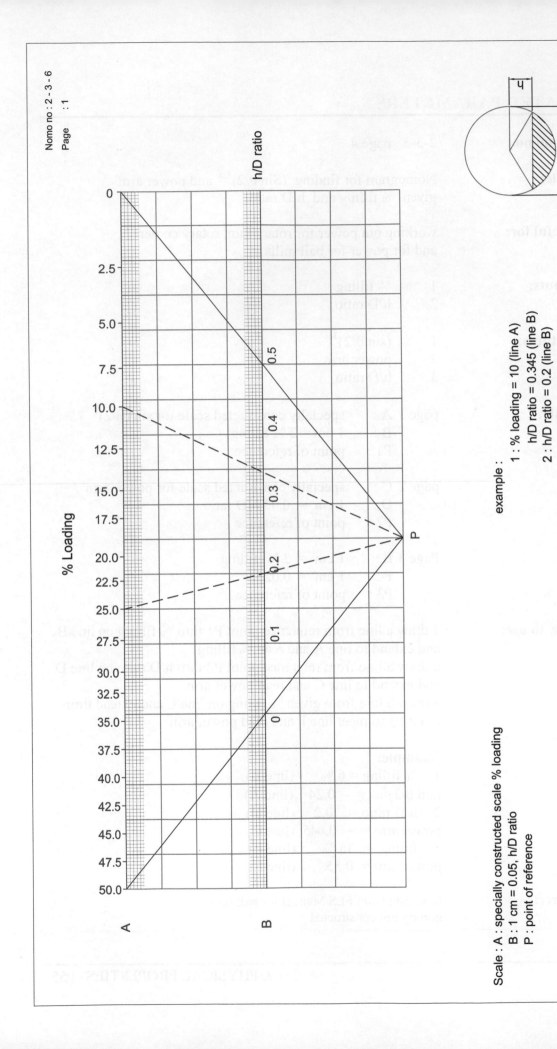

h/D ratio

% Loading

Scale : A : specially constructed scale % loading
        B : 1 cm = 0.05, h/D ratio
        P : point of reference

example :
1 : % loading = 10 (line A)
    h/D ratio = 0.345 (line B)
2 : h/D ratio = 0.2 (line B)
    % loading = 25 (line A)

Nomogram for finding relation between % loading and ratio h/D

Nomo no.:	2-3-6 page 2
Title:	Nomogram for finding relation between % loading and ratio h / D
Useful for:	Working out charge in mills, kilns, coolers etc.
Inputs:	ratio h/D where 'h' = height from center to surface of charge and 'D' = clear dia of mill, kiln etc
Output:	% loading
Scale:	A : specially constructed scale for % loading B : 1 cm = 0.05 ratio h / D P : point of reference
How to use:	Draw a line from point of reference P thru given h / D on line B and extend to meet line A and read % loading

Example:
1   % loading = 10 (line A)
     h / D ratio = 0.345 (line B)
2   h / D ratio = 0.2 (line B)
     % loading ~ 25 (line A)

Source:	FLS Manual for Mills nomogram constructed

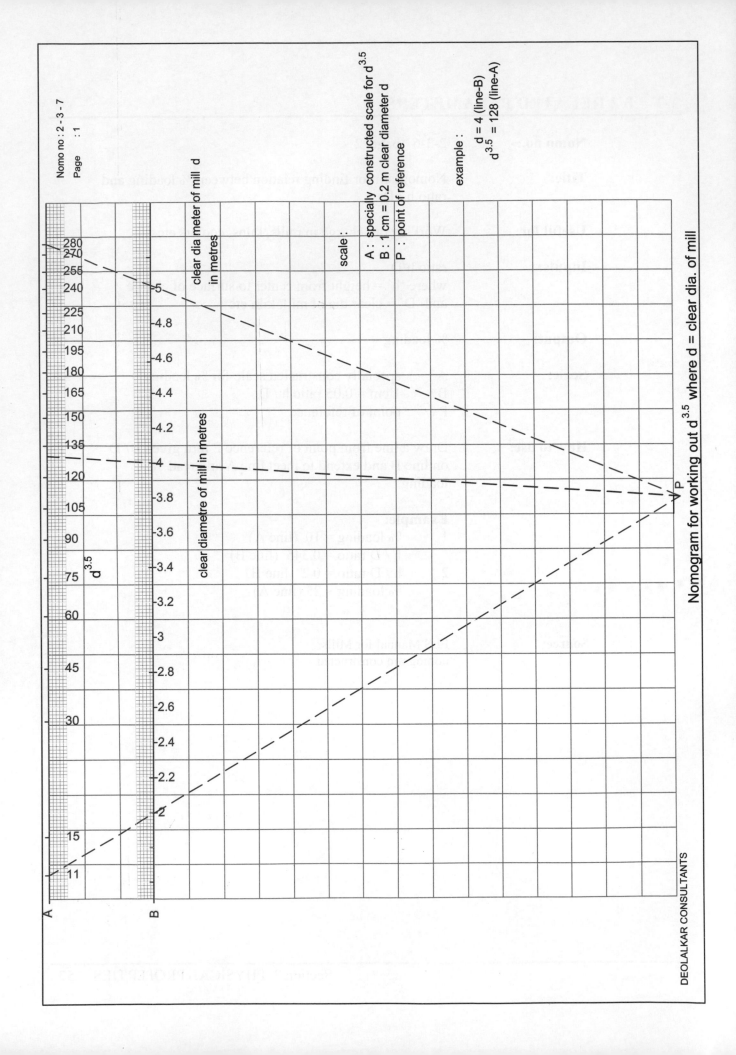

Nomo no : 2 - 3 - 7
Page : 1

clear dia meter of mill d
in metres

scale :

A : specially constructed scale for $d^{3.5}$
B : 1 cm = 0.2 m clear diameter d
P : point of reference

example :
d = 4 (line-B)
$d^{3.5}$ = 128 (line-A)

clear diametre of mill in metres

280
270
255
240
225
210
195
180
165
150
135
120
105
90
75
60
45
30
15
11

$d^{3.5}$

5
4.8
4.6
4.4
4.2
4
3.8
3.6
3.4
3.2
3
2.8
2.6
2.4
2.2
2

A

B

P

Nomogram for working out $d^{3.5}$ where d = clear dia. of mill

DEOLALKAR CONSULTANTS

Nomo no.:	2-3-7 page 2
Title:	Nomogram for working out $d^{3.5}$ where d = clear dia. of mill
Useful for:	Calculating power of ball mills
Inputs:	Diameter 'd' inside liners in metres
Output:	power $d^{3.5}$
Scale:	A : specially constructed scale B : 1 cm = 0.2 metres P : point of reference
How to use:	From point P draw a line thru dia of mill on line B and extend to meet line A and read $d^{3.5}$
	Example: d = 4 on line B $d^{3.5}$ = 128 on line A
Source:	constructed

Nomogram for finding relation between % filling and sin of Ø/2
where Ø/2 is half angle at centre

scale:

A : specially constructed scale % filling (0-20%)
B : sin Ø/2; 1 cm = 0.1
P : point of reference

example:

% filling ≃ 10% (line-A)
sin Ø/2 ≃ 0.74 (line-B)

Nomo no.:	2-3-8 page 2
Title:	Nomogram for finding relation between % filling and sine θ/2, where θ/2 is half angle at centre
Useful for:	calculating power drawn by rotary kilns, coolers and dryers
Inputs:	% filling (in terms of clear cross section of kiln or cooler or dryer)
Output:	sine of corresponding half angle at center θ/2 See sketch
Scale:	A : specially constructed scale for % filling (0-20) B : 1 cm = 0.1 sine of half angle θ/2 P : point of reference
How to use:	from P draw a line to given % filling on line A which cuts line B. Read sine θ/2 on it.

Example:
% filling = 10
Sine of half angle = ~ 0.74

Source:	Duda and others nomogram constructed

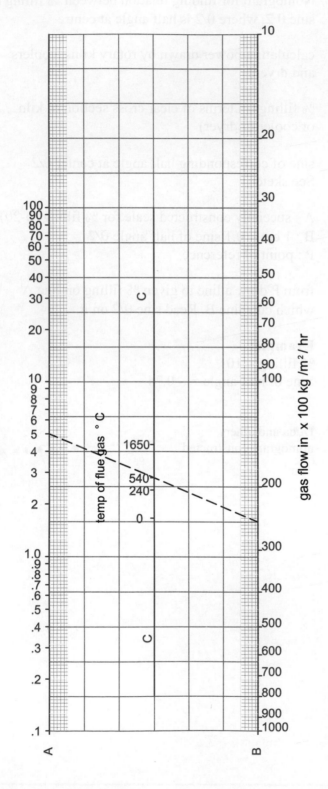

Velocity head in mm wg

temp of flue gas °C

gas flow in x 100 kg /m² /hr

A

B

example : gas flow = 25000 kg /m² / hr.(line B)
          flue gas temp. = 540° C (line C-C)
          then velocity head in mm wg = 5 (line A)

Scale : A : log scale 5 cms = 1-10, 10-100
         B : log scale 10 cms = 10-100, 100-1000

Nomogram for working out velocity head from gas flow in kg /m² /hr at different temperatures

DEOLALKAR CONSULTANTS

Nomo no.:	2-4-1 page 2
Title:	Nomogram for working out velocity head from gas flow in kg/m^2/hr at different temperatures
Useful for:	Design of exhaust gas systems
Inputs:	1 gas flow in hundreds of kg/m^2/hr 2 temperature of flue gas in $^\circ$C
Output:	Velocity head in mm wg
Scale:	A : log scale – 5 cm = 1-10, 10- 100x etc. velocity head B : log scale – 10 cm = 10-100, 100-1000 $\times$ 100 kg/m^2/hr
How to use:	**Example:** Draw a line from given gas flow on line B (25000) thru temperature of flue gas on line C-C (540 $^\circ$C) and extend to line A and read velocity head (5) in mm wg on it
Source:	Base data from 'STEAM' by Babcock nomogram constructed

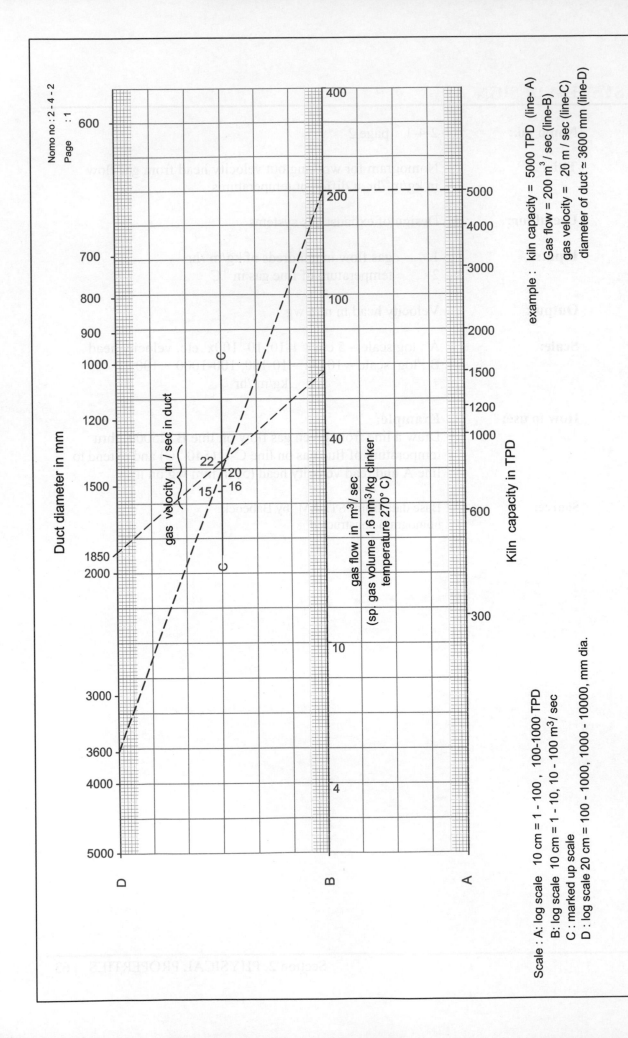

Duct diameter in mm

600

700

800

900

1000

1200

1500

1850
2000

3000

3600

4000

5000

gas velocity m / sec in duct

22
20
15   16

C

C

400

200

100

40

10

4

gas flow in m³/ sec

(sp. gas volume 1.6 nm³/kg clinker
temperature 270° C)

300

600

1000
1200

1500

2000

3000

4000

5000

Kiln capacity in TPD

example :   kiln capacity = 5000 TPD   (line- A)
            Gas flow = 200 m³ / sec (line-B)
            gas velocity = 20 m / sec (line-C)
            diameter of duct ≃ 3600 mm (line-D)

D

B

A

Scale : A: log scale 10 cm = 1 - 100 , 100-1000 TPD
        B: log scale 10 cm = 1 - 10, 10 - 100 m³/sec
        C : marked up scale
        D : log scale 20 cm = 100 - 1000, 1000 - 10000, mm dia.

Nomogram for sizing down comer duct for different kiln capacities and velocities in duct

DEOLALKAR CONSULTANTS

Nomo no.:	2-4-2 page 2
Title:	Nomogram for sizing down comer duct for different kiln capacities and velocities in duct
Useful for:	Quick sizing of ducts when gas flows are known

Inputs:	1	kiln capacity in tpd 0- 5000 tpd
	2	sp. gas volume = 1.6 nm^3/kg clinker
	3	exhaust temperature = 270 $^\circ$ C (6 stage prheater)
	4	draft at outlet of prheater - 500 mmwg
	5	velocities in duct 16-22 m/sec.

Output:	Duct diameter in mm
	Note: If quantum of tertiary air is known , same nomogram can be used to calculate clear diameter of tertiary duct also.
Scale:	A : log scale : 10 cm = 1-10, 10-100 etc
	B : log scale : 10 cm = 1-10, 10-100 etc.
	C : marked up scale
	D : log scale : 20 cm = 100-1000, 1000-10000 etc.
How to use:	From given capacity of kiln on line D, find gas volume in m^3/sec on line C.
	From this point draw a line through appropriate velocity marked on line C and extend to meet line A and read dia of duct on it in mm

Example:
kiln capacity   = 5000 tpd (line A)
Gas volume   =   ~ 200 m^3/sec (line B)
Velocity   =   20 m/sec   (line C)
Duct diameter = ~ 3600   mm (line D)
Given gas volume and desired velocity in duct, size
of duct can be calculated for conveying tertiary air
and for kilns of any capacities
assuming that tertiary air volume is 60 m^3/sec, and
velocity in duct is 22   m/sec then clear size of tertiary
duct = ~ 1850 mm

Source:	constructed

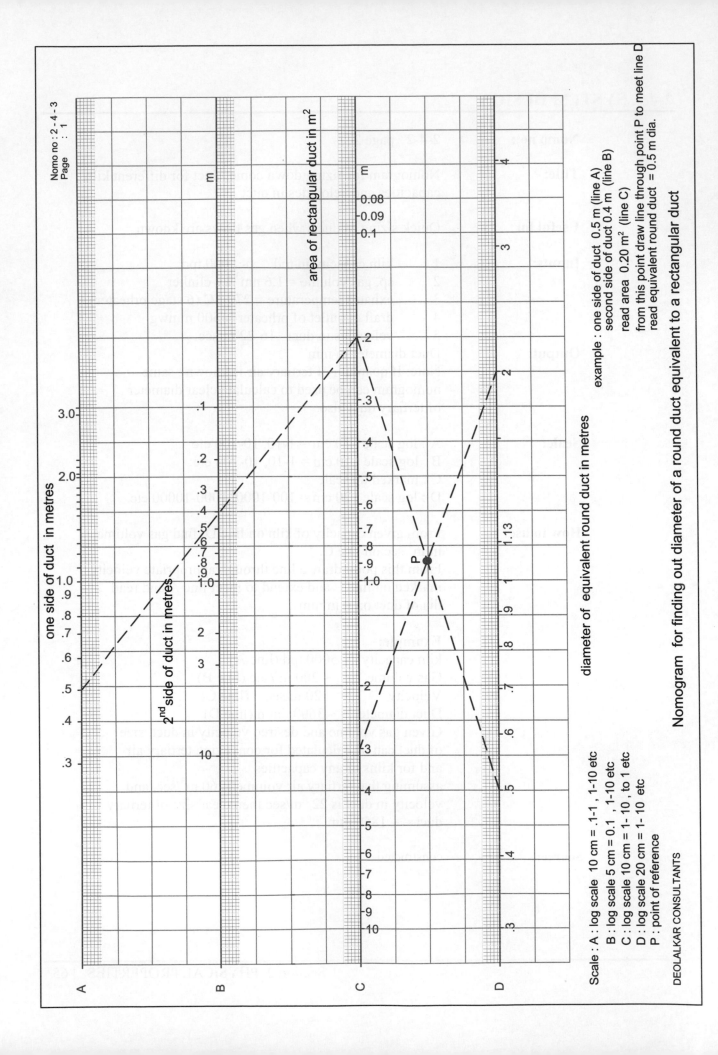

Nomo no : 2 - 4 - 3
Page : 1

A : one side of duct in metres
3.0
2.0
1.0 .9 .8 .7 .6 .5 .4 .3

B : 2nd side of duct in metres
.1
.2
.3
.4
.5 .6 .7 .8 .9 1.0
2
3
10

C : area of rectangular duct in m²

m
0.08
0.09
0.1

.2
.3
.4
.5
.6
.7
.8
.9 1.0
2
3
4
5
6
7
8
9
10

D : diameter of equivalent round duct in metres

m
4
3
2
1.13
1
.9
.8
.7
.6
.5
.4
.3

P

example : one side of duct 0.5 m (line A)
second side of duct 0.4 m (line B)
read area 0.20 m² (line C)
from this point draw line through point P to meet line D
read equivalent round duct = 0.5 m dia.

Scale : A : log scale 10 cm = .1-1 , 1-10 etc
B : log scale 5 cm = 0.1 , 1-10 etc
C : log scale 10 cm = 1- 10 , to 1 etc
D : log scale 20 cm = 1- 10 etc
P : point of reference

Nomogram for finding out diameter of a round duct equivalent to a rectangular duct

DEOLALKAR CONSULTANTS

Nomo no.:	2-4-3 page 2
Title:	Nomogram for finding diameter of a round duct equivalent to a rectangular duct
Useful for:	For calculating pressure loss in gas/air stream with rectangular ducts
Inputs:	Sides of rectangular duct in m
Output:	Diameter of equivalent round duct in m
Scale:	A : one side rect. duct : log scale 10 cm = 0.1-1 etc
	B : second side rect. duct : log scale 5 cm = 0.1-1 etc
	C : area of rect. duct : log scale 10 cm = 1-10 etc.
	D : equivalent round duct : log scale 20 cm = 1-10 etc.
How to use:	Draw a line from 1st side of rect. duct on line A thru second side of rect. duct on line B, to meet line C and read area of duct. From this point draw line thru point P to meet line D and read dia. of equivalent round duct

Example:
one side of rect. duct = 0.5 m (line A)
second side   = 0.4 m (line B)
area of rect. duct   = 0.2 m^2 (line C)
dia. of equivalent round duct = ~ 0.5 m (line D)

Source:	constructed

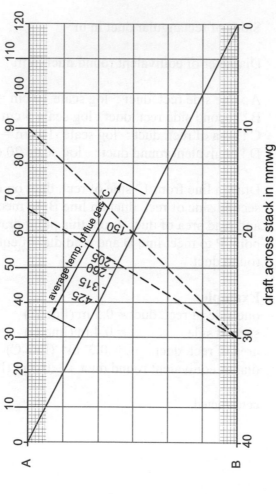

Height of stack in metres

draft across stack in mmwg

average temp. of flue gas °C

Scale :
A : 1 cm = 10 m
B : 3 cm = 10 mmwg

example :
draft across stack = 30 mm (line B)
average temp of flue gas = 150 °C (0-0)
height of stack = 90 m (line A)

Nomogram for finding height of stack for flue gases of different temperatures
and for different drafts across the stack

Nomo no.:	2-4-4 page 2
Title:	Nomogram for finding height of stack for flue gases of different temperatures and for different drafts across the stack
Useful for:	Designing stacks
Inputs:	1 draft across stack mm wg 2 temperature of flue gas ° C
Output:	Height of stack in metres
Scale:	A : 1 cm = 10 m B : 3 cm = 10 mmwg
How to use:	Draw a line from draft on line B thru flue gas temperature on line 0-0 and extend to line A and read height of stack in metres

Example:
draft across stack = 30 mm (line B)
temp. of flue gas = 150 ° C (line 0-0)
height of stack   = ~ 90 m

Source:	base data from Babcock book 'Steam' nomogram constructed

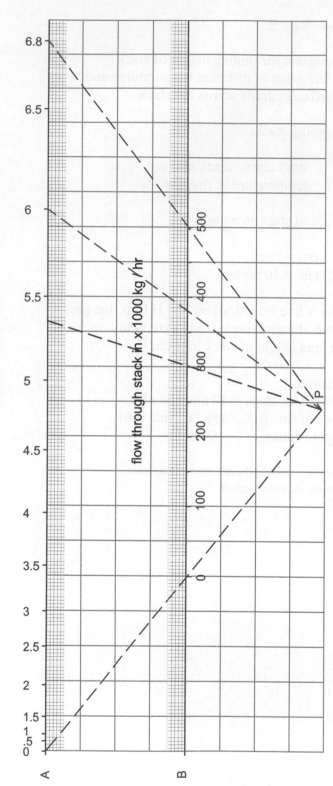

stack diameter in metres

flow through stack in x 1000 kg / hr

example:

flow : 300000 kg / hr (line B)
dia of stack = 5.35 m (line A)

scale :

A : specially constructed for diameter of stack
B : 1 cm = 50.000 kg / hr

Nomogram for finding diameters of stacks with natural draft according to weight of flue gas in kg / hr

DEOLALKAR CONSULTANTS

Nomo no.:	2-4-5 page 2
Title:	Nomogram for finding diameters of stacks with natural draft according to weight of flue gas in kg/hr
Useful for:	Quickly working out dimensions of stack with the help of nomo no 2-4-6
Inputs:	Flow of flue gas in kg/hr
Output:	diameter of stack in metres
Scale:	A : specially constructed scale for diameter of stack B : 1 cm = 50 × 1000 kg/hr
How to use:	Draw a line from reference point P thru quantity of flue gas on line B and extcnd to line A to read dia. of stack in metres

Example:
flue gas quantity = 300000 kg/hr (line B)
diameter of stack = ~ 5.35 m

Source:	base data from Babcock book 'Steam' nomogram constructed

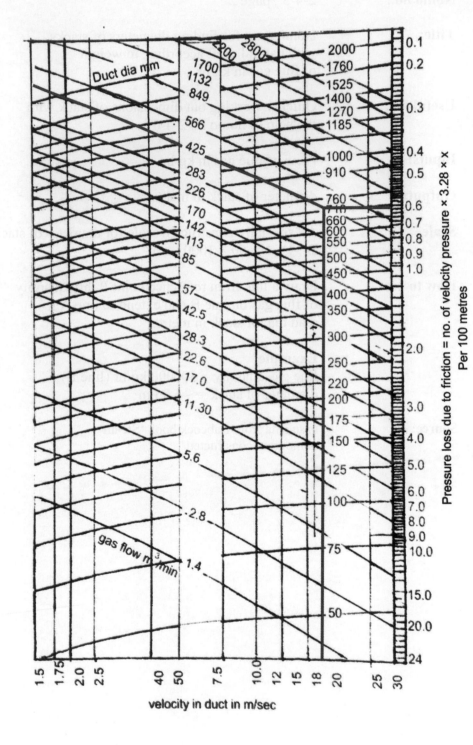

Pressure loss due to friction = no. of velocity pressure × 3.28 × x
Per 100 metres

velocity in duct in m/sec

Nomo no.:	2-4-6 page 2
Title:	Nomogram for working out pressure loss in pipes and ducts due to flow of gases
Useful for:	Working out losses in pressure in various legs of a system and to arrive at specifications of fans.
Inputs:	Flow of gas in m^3/ min., its temp. in $°C$, altitude in metres and velocity of gas in m/sec. Length of duct – straight in metres
Output:	pr. loss in a straight duct of 100 ms length in mm wg.
How to use:	Follow the curve of gas flow in m^3/ min. till it crosses the line showing velocity in the duct. From the point of intersection, draw a horizontal line till it meets the scale on the right Read the scale say 'x' Pr. loss in mm in a 30 m long st. pipe is given by : 'x' × velocity pressure × 3.28 mm wc
Example:	Let flow of gas = 425 m^3/ min, temp. of gas be $100°$ C and the altitude at site = 300 metres and velocity thru duct be 18 m/sec. Pt . of intersection x = ~ 0.65 Velocity head@ 18 m/sec = $18^2/19.62 = 16.51$ Density of air @ $100°$ C and 300 metres altitude = ~ 0.91 kg/m^3 from nomogram 2-1-6 velocity pressure for velocity of 18 m/sec and density of 0.91 kg/m^3, = ~ 15 mm from nomogram 2-1-7 therefore pr. loss in duct of 100 m length = 15 × 0.65 × 3.28 = 32 mm wc. Substitute actual equivalent length of pipe to obtain actual pr. loss in the duct.
Source:	Engineering Memoranda of ACC Babcock Ltd. See also annexures 1 & 2 of Chapter 43, Section 6 of the Book 'Handbook for Designing Cement Plants' by S.P. Deolalkar

Section 3 PROCESS

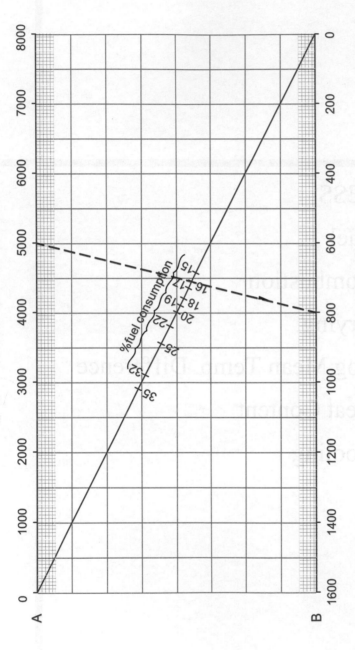

Useful calorific value of coal kcal/kg

sp.fuel consumption in kcal/kg clinker

example : useful calorific value of coal = 5000 kcal/kg; % fuel consumption = 16

∴   sp. fuel consumption = 800 kcal/kg clinker

(knowing sp. fuel consumption & calorific value of coal
coal consumption in % can be found)

Nomogram for working out sp. fuel consumption in kcal/kg clinker from % fuel
consumption & vice versa

Scale : A: 1 cm = 500 kcal/kg
        B: 1 cm = 100 kcal/kg clinker

DEOLALKAR CONSULTANTS

Nomo no.:	3-1-1 page 2
Title:	Nomogram for working out sp. fuel consumption in kcal/kg of clinker from % fuel consumption and vice versa
Useful for:	All process calculations, heat balances etc.
Inputs:	1 useful calorific value of coal /oil in kcal/kg
	2 % fuel consumed to produce 1kg clinker alternatively sp. Fuel consumption in kcal/kg clinker
Output:	sp. Fuel consumption in kcal/kg clinker Alternatively % fuel consumed per kg clinker
Scale:	A : 1cm = 500 kcal/kg of fuel B : 1cm = 100 kcal/kg clinker
How to use:	Draw a line from given calorific value of coal on line A through % fuel consumption on line 0-0 to meet line B and read sp. Fuel consumption in kcal/kg clinker. Can also be used to find % fuel consumption by joining pertinent points on lines A and B and read % fuel on Line 0-0

Example:
useful calorific value of coal – 5000 kcal/kg
                        (line A)
% fuel consumption - 16 % (line 0-0)
sp. fuel consumption – 800 kcal/kg (line B)

Source:	constructed

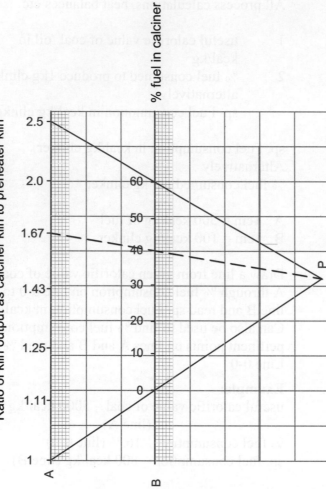

% fuel in calciner

Ratio of kiln output as calciner kiln to preheater kiln

Scale:

A : specially constructed scale for ratio

B : 1 cm = 10 % fuel

Example :

% fuel in calciner = 40 (line B)

ratio of increase in kiln capacity = 1.67 (line A)

Nomogram for several variables dependent on % fuel fired in calciner

Nomo no.:	3-1-2 page 2
Title:	Nomogram for several variables dependent on % fuel fired in calciner
Useful for:	quick assessment of capacities of preheater and calciner kilns

Note : % calcinations and L.O.I are sort of reciprocal and both are useful for specific applications
When fuel in calciner is 0 %, it is a preheater kiln. Ratio is 1 and sp. output is 1.7 tpd/m^3 respectively
See pages 1 & 4

Inputs: % fuel fired in calciner

Outputs:

1)   % degree of calcination at kiln inlet
2)   increase in kiln capacity expressed as ratio
3)   specific output in tpd/m^3

Scale:

Nomo no: 3-1-2-1

increase in kiln capacity
A : specially constructed scale for increase in
    Kiln capacity
B : 1 cm = 10 %
P : point of reference

Nomo no 3-1-2-3

1   % Degree of calcination
C : specially constructed scale for % calcination
D : 1 cm = 10 % fuel
P : point of reference

Nomo no 3-1-2-5

2   sp. output in tpd/m^3
C : specially constructed scale for sp. output
D : 1 cm = 10 %
P : point of reference

How to use:

In each of these three cases, draw a line from the respective points of reference P to the % fuel in calciner on lines B, D, F and extend to meet line A, C, E and read on them respective entities as applicable

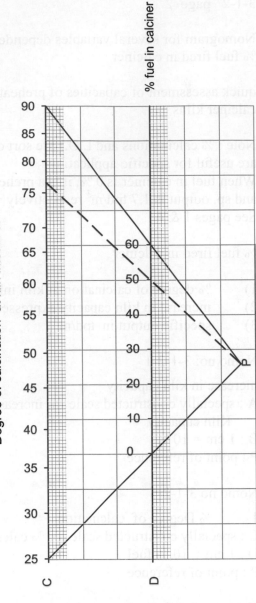

% fuel in calciner

Degree of calcination at kiln inlet in %

C

D

P

Example :

% fuel in calciner = 50 % (line D) : draw line thru P and this point

to meet line C and read % calcination at kiln inlet ~ 77%

Nomogram for several variables dependent on % fuel fired in calciner

Scale:

C : specially constructed scale for
    % calcination

D : 1 cm = 10% fuel

P : Point of reference

DEOLALKAR CONSULTANTS

Nomo no.:   3-1-2 page 4

Example:

1   ratio of increase in capacity of kiln
        % fuel in calciner 40 % (line B),
        increase in kiln capacity 1.67 times (line A)

2   % degree calcination at kiln inlet
        Thru 50 % fuel (line D) and extend to
        (line C) and read degree of calcinations as 77 %

3   specific output in tpd / m^3
        fuel in calciner 50 % (line F)
        sp. output of kiln = ~ 3.4 tpd/m^3 (line E)

Source: base data from Onoda Manual
                               nomogram constructed

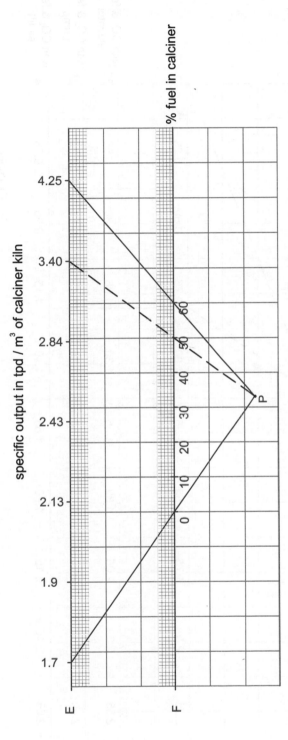

specific output in tpd / m³ of calciner kiln

% fuel in calciner

Scale:

E : specially constructed scale for sp. output
F : 1 cm = 10 % fuel
P : Point of reference

example :

% fuel in = 50% (line F)
specific output of calciner kiln = 3.4 tpd / m³ (line E)

Nomogram for several variables dependent on % fuel fired in calciner

DEOLALKAR CONSULTANTS

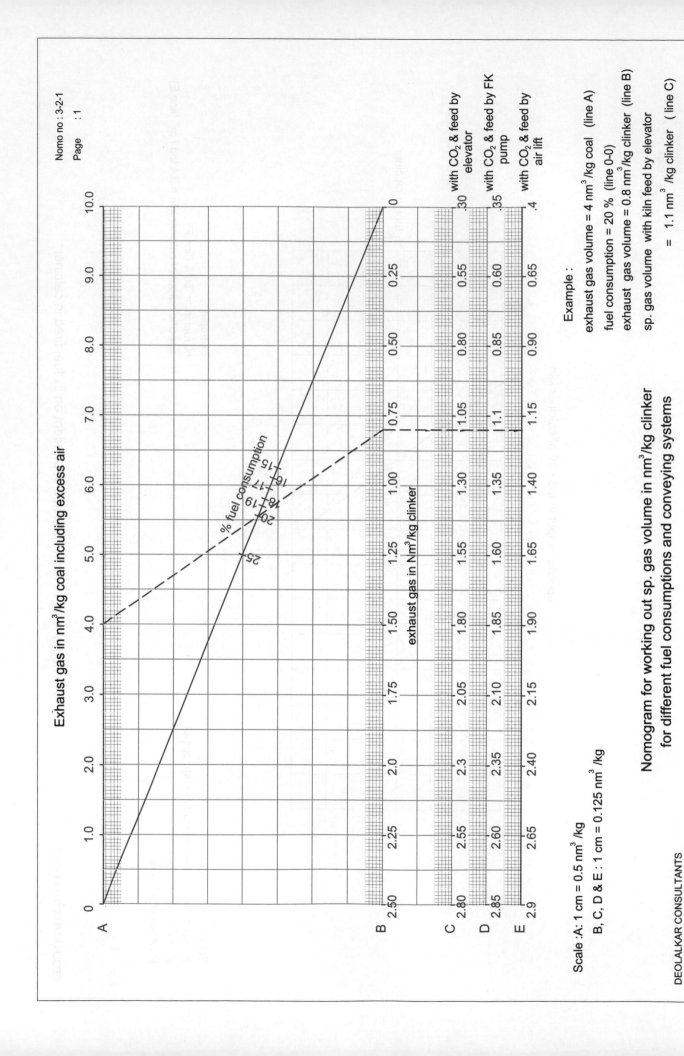

Nomo no : 3-2-1
Page : 1

Exhaust gas in nm³/kg coal including excess air

% fuel consumption

exhaust gas in Nm³/kg clinker

with CO₂ & feed by elevator
with CO₂ & feed by FK pump
with CO₂ & feed by air lift

Example :
exhaust gas volume = 4 nm³/kg coal   (line A)
fuel consumption = 20 % (line 0-0)
exhaust gas volume = 0.8 nm³/kg clinker (line B)
sp. gas volume with kiln feed by elevator
= 1.1 nm³ /kg clinker (line C)

Scale :A: 1 cm = 0.5 nm³ /kg
B, C, D & E : 1 cm = 0.125 nm³ /kg

Nomogram for working out sp. gas volume in nm³/kg clinker
for different fuel consumptions and conveying systems

DEOLALKAR CONSULTANTS

Nomo no.:	3-2-1 page 2
Title:	Nomogram for working out sp. gas volume in nm^3/kg clinker for different fuel consumptions and conveying systems
Useful for:	Working out gas volume at the exit of preheater-knowing exhaust gas volume including excess air -for different fuel consumptions and conveying systems

Inputs:

1   exhaust gas volume including excess air in nm^3/kg coal
2   % coal consumption
3   CO_2 released $= 0.3$ nm^3/kg clinker
4   conveying air which is:
        0.0 nm^3/kg clinker for elevator
        0.05 nm^3/kg clinker for fk pump
        0.1 nm^3/kg clinker for air lift

Output:	Sp. gas volume in nm^3/kg clinker
Scale:	A : 1 cm $= 0.5$ nm^3/kg B : 1 cm $= 0.125$ nm^3/kg C, D & E, 1 cm $= 0.125$ nm^3/kg
How to use:	Draw a line from given exhaust gas volume on line A through % fuel consumption on line 0-0 to meet line B. Read sp. gas volume corresponding to the conveying system on line C, D or E

Example:

exhaust gas	$= 4$ nm^3/kg coal (line A)
% fuel consumption	$= 20$ (line 0-0)
exhaust gas in nm^3/kg clinker	$= 0.8$ (line B)
conveying system	$=$ elevator
sp. gas volume	$= 1.1$ nm^3/kg clinker (line C)

Source:	Constructed

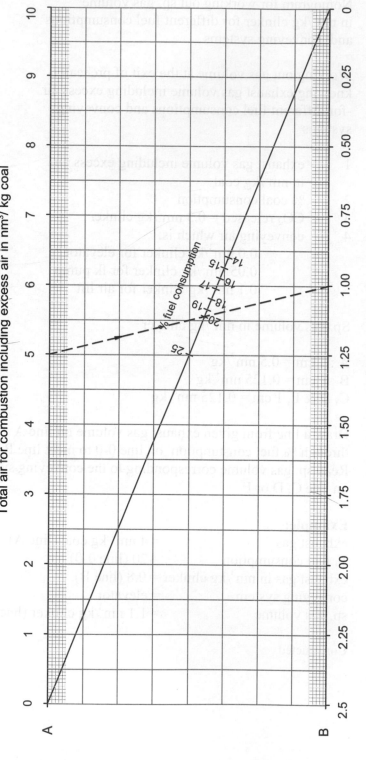

Total air for combustion including excess air in nm³ / kg coal

Total air for combustion in nm³ / kg clinker

% fuel consumption

Example :

Total air for combustion = 5 nm³ / kg coal (line A)
% fuel consumption = 20 (line 0-0)
Total air for combustion in nm³ / kg clinker = 1 (line B)

Nomogram for total air for combustion in nm³ / kg clinker
for different rates of fuel consumption

Scale : A: 1 cm = 0.5 nm³ / kg coal
        B: 1 cm = 0.125 nm³ / kg clinker

Nomo no.:	3-2-2 page 2
Title:	Nomogram for working out total air for combustion in nm^3/kg clinker for different rates of fuel consumption
Useful for:	Working out quantities of air for combustion in nm^3/kg clinker
Inputs:	1 total air for combustion in nm^3/kg coal including excess air 2 % coal consumption
Output:	Total air for combustion in nm^3/kg clinker
Scale:	A : 1 cm = 0.5 nm^3/kg coal B : 1 cm = 0.125 nm^3/kg clinker
How to use:	Draw a line from total air for combustion on line A through % fuel consumption on line 0-0 to meet line B and read total air for combustion in nm^3/kg clinker **Example:** total air for combustion = 5 nm^3/kg (line A) % fuel consumption = 20 % (line 0-0) air for combustion = 1.0 nm^3/kg clinker (line B)
Source:	constructed

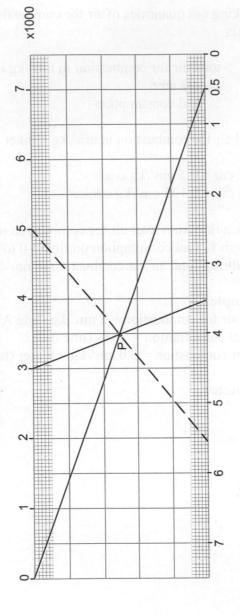

x1000

Useful calorific value of coal kcal/kg

Theoretical air for combustion nm³ /kg coal

A

B

Example:

calorific value of coal : 5000 kcal/kg (line A)

Theoretical air for combustion : 5.55 nm³ /kg coal (line B)

Nomogram for working out theoretical air for combustion for different calorific values of coal

Scale :

A : 2 cms = 1000 kcal/kg

B : 2 cms = 1nm³ /kg

Nomo no.:	3-2-3 page 2
Title:	Nomogram for working out total air for combustion with different excess air factors and different calorific values of coal
Useful for:	working out quantities of total air for combustion and excess air for calculating exhaust gas volumes after preheater

Inputs:

1   useful calorific value of coal in kcal/kg
2   % excess air
3   excess air factors

Combustion air = $(1.01 \times H/1000 + 0.5)$ nm^3/kg coal
where H = calorific value of coal in kcal/kg

Outputs:

1   Theoretical air for combustion for different calorific values of coal
2   quantity of excess air nm^3/kg for different percentages of excess air
3   total air for combustion including excess air

Scale:

Page 1
A : 1 cm = 500 kcal/kg
B : 1 cm = 0.5   nm^3/kg
Page 3
A : 1 cm = 500 kcal/kg
B : 1 cm = 0.25 nm^3/kg
C : 1 cm = 500 kcal/kg
D : 1 cm = 0.5 nm^3/kg coal

How to use:

1   Theoretical air for combustion 3-2-3- page 1
draw a line from given calorific value of coal on line A through point P and extend it to meet line B and read theoretical air for combustion in nm^3

2   Excess air for combustion for different percentage of excess air   3-2-3- page 3
draw a line from given calorific value of coal on line A through desired % excess air on line 0-0 to meet line B and read quantity of excess air in nm^3/kg coal

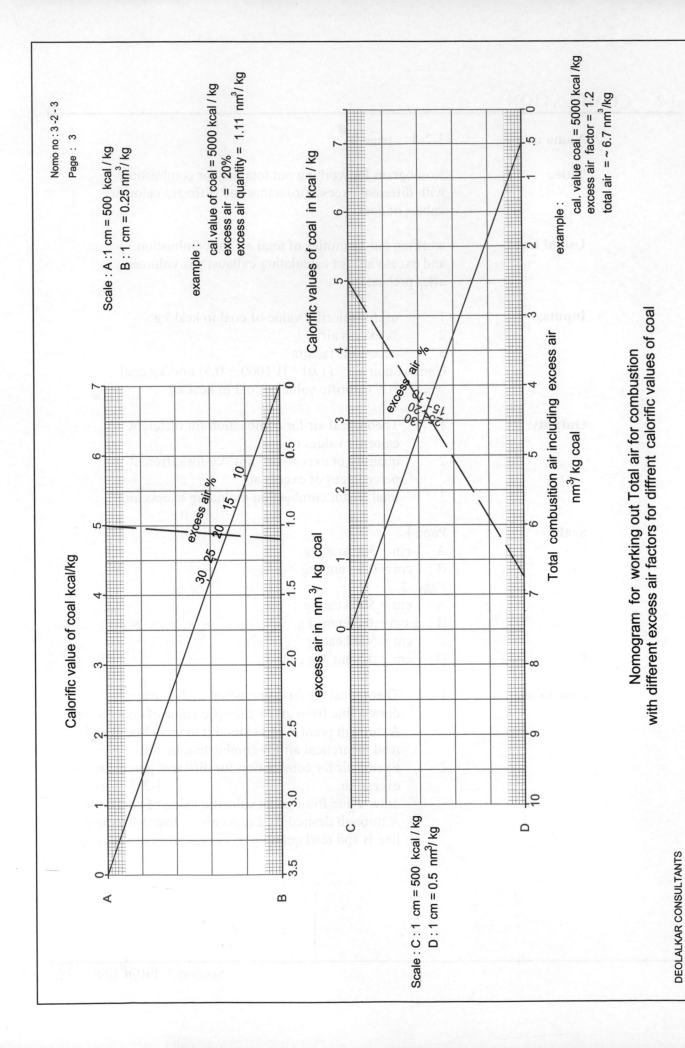

Nomo no : 3 - 2 - 3

Page : 3

Scale : A : 1 cm = 500 kcal / kg
B : 1 cm = 0.25 nm³ / kg

example :

cal.value of coal = 5000 kcal / kg
excess air = 20%
excess air quantity = 1.11 nm³ / kg

Calorific value of coal kcal/kg

excess air %

excess air in nm³ / kg coal

Scale : C : 1 cm = 500 kcal / kg
D : 1 cm = 0.5 nm³/kg

Calorific values of coal in kcal / kg

example :

cal. value coal = 5000 kcal /kg
excess air factor = 1.2
total air = ~ 6.7 nm³/kg

excess air %

Total combustion air including excess air
nm³/ kg coal

Nomogram for working out Total air for combustion
with different excess air factors for diffrent calorific values of coal

DEOLALKAR CONSULTANTS

Nomo no.:   3-2-3 page 4

3   Total air for combustion including excess air
                    3-2-3 page 3
            Draw a line from given calorific value of coal on

            line C thru desired % of excess air on line 0-0 to
            meet line D and read total air for combustion
            in nm^3/kg coal

	excess air	
%		factor
10		1.1
15		1.15
20		1.2
30		1.3

Example:
calorific value of coal   = 5000 cal/kg
theoretical air for combustion = 5.55 nm^3/kg coal
from page 3, lines A and B
calorific value of coal   = 5000kcal/kg and
% excess air 20.
Quantity of excess air   = 1.11 nm^3/kg coal
From page 3, lines C and D
Total air for combustion with 20 % excess air
~ 6.7 nm^3/kg coal from page 3, lines C and D

Source: formula for air for combustion from Onoda /Mitsubishi
            manuals
            nomogram constructed

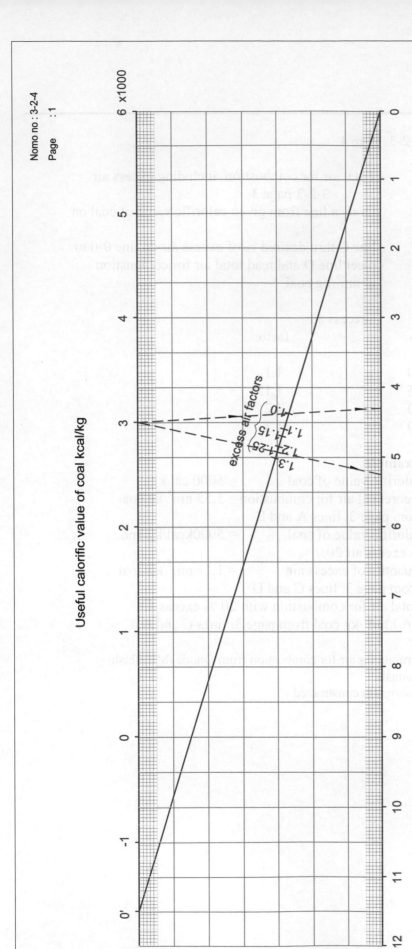

Useful calorific value of coal kcal/kg

Product of combustion & exhaust gas volume
for different quantities of excess air in nm³/kg coal

excess air factors

0-0'
1.1, 1.15
1.2, 1.25
1.3

Nomo no : 3-2-4
Page   : 1

example :
calorific value of coal = 3000 kcal/kg (line A)
excess air factor = 1 (line 0-0')
products of combustion = 4.32 nm³/kg (line B)
excess air factor = 1.25 (line 0-0')
exhaust gasvolume = ~ 5.2 nm³/kg coal (line B)

Nomogram for working out products of combustion and exhaust gas volumes for different
calorific values of coal and quantities of excess air

Scale : A: 3 cm = 1000 kcal/kg
        B: 1 cm = 0.5 nm³ / kg

DEOLALKAR CONSULTANTS

Nomo no.:	3-2-4 page 2
Title:	Nomogram for working out products of combustion and exhaust gas volume for different calorific values of coal and different quantities of excess air
Useful for:	working out products of combustion and exhaust gas volumes for different calorific values of coal and quantities of excess air
Inputs:	1 useful calorific values of coal in kcal/kg 2 % excess air Formula for product of combustion: $V = 0.89 \times H/1000 + 1.65$ nm^3/kg coal Exhaust gas = product of combustion + excess air
Outputs:	1 products of combustion in nm^3/kg coal 2 exhaust gas volume in nm^3/kg coal
Scale:	A : 3 cm = 1000 kcal/kg B : 1 cm = 0.5 nm^3/kg coal
How to use:	Draw a line from calorific value of coal on line A through 0 % excess air on line 0'-0 to meet line B and read products of combustion in nm^3/kg coal Draw similarly line through different % s of excess air to read exhaust gas volumes in nm^3/kg coal.

Example:
calorific value of coal = 3000 kcl/kg (line A)
Excess air : 0 % (line 0'-0)
Product of combustion = ~ 4.32 nm^3/kg coal
Excess air 25 % (line 0'-0)
Product of combustion = ~ 5.2 nm^3/kg coal
      (line B)

Source:	formula for product of combustion from Onoda/Mitsubishi manuals nomogram constructed

Total combustion air thru kiln & calciner including excess air nm³/kg clinker

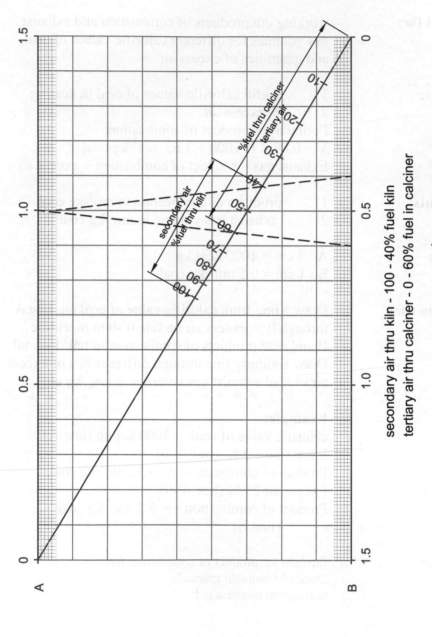

secondary air thru kiln - 100 - 40% fuel kiln
tertiary air thru calciner - 0 - 60% fuel in calciner

example : total combustion air = 1.0³nm / kg (line A)
              % fuel in calciner = 60 % (line 0-0)
              air through calciner 0.6 nm³/ kg (line B)
              air through kiln = 0.4 nm³ /kg (line B)

**Nomogram for finding out distribution of secondary & tertiary airs
for different proportions of fuel in kiln & calciner**

Scale : A : 1 cm = 0.1 nm³ / kg
          B : 1 cm = 0.1 nm³ / kg

DEOLALKAR CONSULTANTS

Nomo no.:	3-2-5 page 2
Title:	Nomogram for working out distribution of secondary and tertiary airs for different proportions of fuel in kiln and calciner
Useful for:	working out quantity of secondary air through kiln and tertiary air through calciner from given total air for combustion through kiln and calciner in nm^3/kg clinker
Inputs:	1 Total quantity of secondary and tertiary air for combustion including excess air in nm^3/kg clinker 2 % fuel fired in kiln and calciner
Output:	1 secondary air through kiln in nm^3/kg clinker 2 tertiary air through calciner
Scale:	A : 1 cm = 0.1 nm^3/kg B : 1 cm = 0.1 nm^3/kg
How to use:	Draw a line from given total air for combustion on line A through % fuel in kiln and calciner on line 0-0 and extend to meet line B and read secondary or tertiary air as the case may be in nm^3/kg clinker

Example:
total air for combustion – 1 nm^3/kg clinker (line A)
fuel in calciner   - 60 % (line 0-0)
air thru kiln   - 0.4 nm^3/kg clinker (line B)
air thru calciner   - 0.6 nm^3/kg clinker (line B)

Source:	constructed

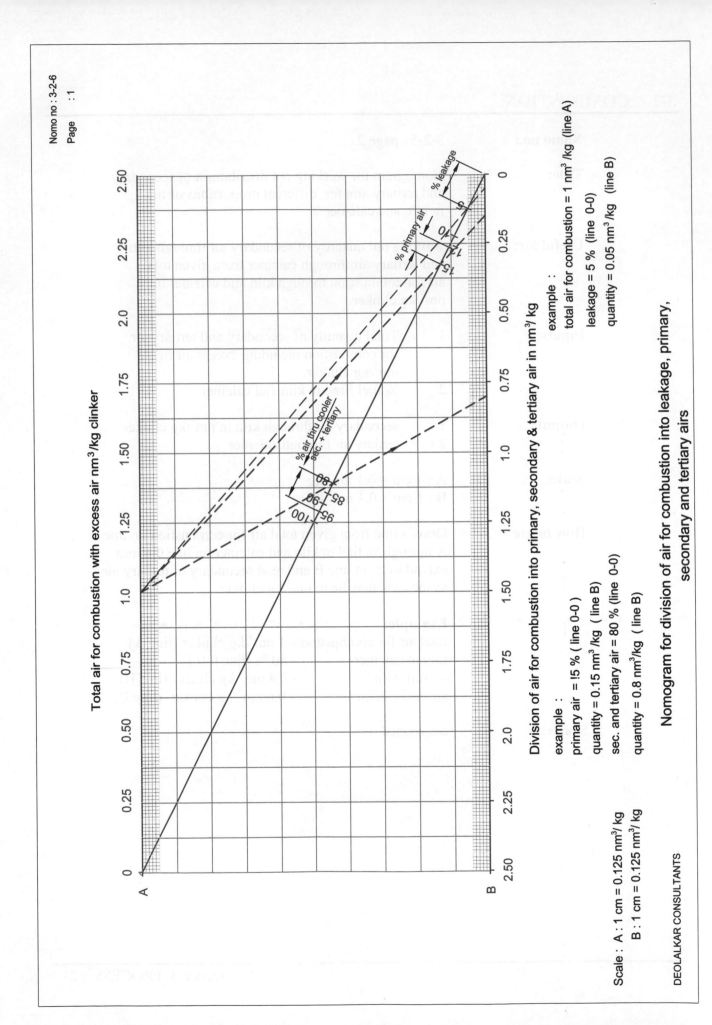

Total air for combustion with excess air nm³/kg clinker

Division of air for combustion into primary, secondary & tertiary air in nm³/ kg

example :
total air for combustion = 1 nm³/kg (line A)
leakage = 5 % (line 0-0)
quantity = 0.05 nm³/kg   (line B)

example :
primary air = !5 % (line 0-0)
quantity = 0.15 nm³/kg (line B)
sec. and tertiary air = 80 % (line 0-0)
quantity = 0.8 nm³/kg (line B)

Nomogram for division of air for combustion into leakage, primary, secondary and tertiary airs

Scale : A : 1 cm = 0.125 nm³/ kg
         B : 1 cm = 0.125 nm³/ kg

DEOLALKAR CONSULTANTS

Nomo no.:	3-2-6 page 2
Title:	Nomogram for division of air for combustion into leakage, primary, secondary and tertiary airs
Useful for:	knowing quantities of primary air, secondary and tertiary airs, leakage etc. from given quantity of total air for combustion including excess air

Inputs:

1   Total air for combustion including excess air
2   percentages of leakage, primary, secondary and tertiary airs

Outputs:

1   quantities of leakage, primary and secondary and tertiary airs in nm^3/kg clinker

Scale:

A :   1 cm = 0.125 nm^3/kg
B :   1 cm = 0.125 nm^3/kg

How to use: draw a line from quantum of total air for combustion on line A through appropriate % ages on line 0-0 and extend to line B and read appropriate quantity of leakage or primary or secondary or tertiary air in nm^3/kg per clinker

Example:
total air for combustion = 1 nm^3/kg clinker (line A)
leakage @ - 5 % (line 0-0)
                 = 0.05 nm^3/kg clinker (line B)
primary air @ 15 % (line 0-0)
                 = 0.15 nm^3/kg clinker (line B)
air thru kiln and calciner @ 80 % (line 0-0)
(secondary and tertiary air)
                 = 0.8 nm^3/kg clinker (line B)

Source: constructed

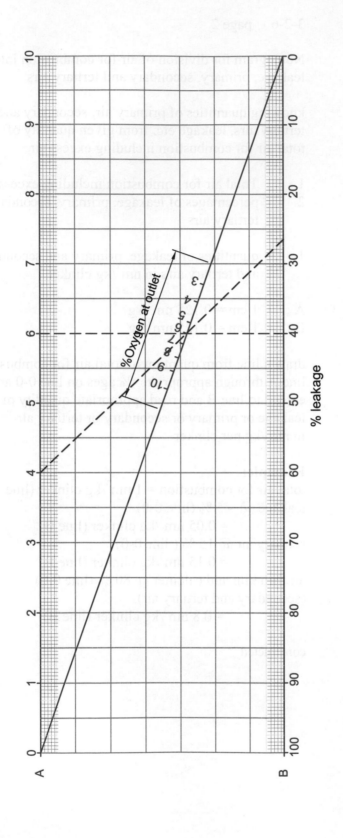

Difference in % Oxygen between outlet & Inlet

% Oxygen at outlet

% leakage

Scale : A : 1 cm = 0.5% difference in % Oxygen between outlet & Inlet

B : 1 cm = 5%   leakage

Example : difference in % Oxygen = 4 - line A

% Oxygen @ outlet = 6 - line 0-0

% leakage ≃ 26 (line B)

Nomogram for finding out % leakage in system
given Oxygen content in gases at inlet & outlet of the system

Nomo no.:	3-2-7 page 2
Title:	Nomogram for finding out % leakage in the system given Oxygen content in gases at inlet and outlet of the system
Useful for:	Working out gas volumes at exit of system and for taking corrective action
Inputs:	1 % Oxygen at inlet 2 % Oxygen at outlet
Output:	% leakage
Scale:	A : 1 cm = 0.5 % difference in Oxygen between inlet and outlet B : 1 cm = 5 % system leakage

Formula:

$$\% \text{ leakage air} = \frac{\text{difference between \% oxygen at outlet \& inlet}}{21 - \% \text{ oxygen at outlet}}$$

How to use: Line A shows difference between Oxygen at inlet and outlet of system. From given difference on line A draw a line through point showing Oxygen at outlet on line 0-0 and extend to meet line B. Read % leakage on it

Example:
difference in % Oxygen at outlet & inlet = 4 % (line A)
Oxygen at outlet = 6 % (line 0-0)
% leakage in system $\simeq$ 26 % (line B)

Source: basic equation from Cement Managers' Handbook nomogram constructed

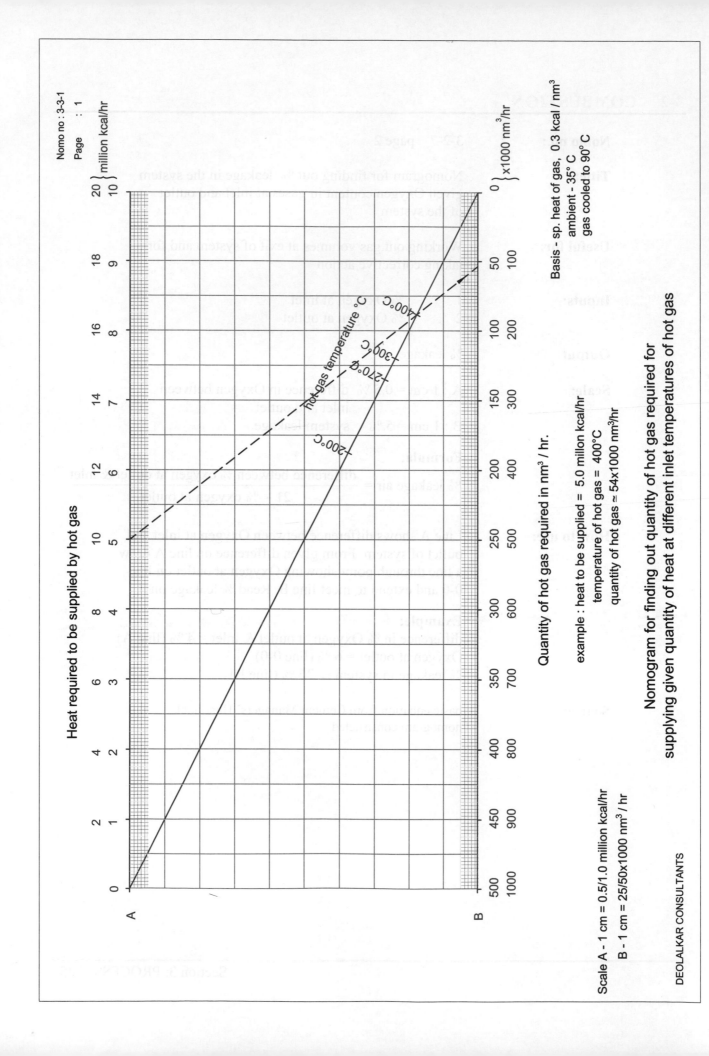

Nomo no : 3-3-1
Page : 1

Heat required to be supplied by hot gas

million kcal/hr

Quantity of hot gas required in nm³ / hr.

example : heat to be supplied = 5.0 millon kcal/hr
temperature of hot gas = 400°C
quantity of hot gas ≈ 54x1000 nm³/hr

Basis - sp. heat of gas, 0.3 kcal / nm³
ambient - 35° C
gas cooled to 90°C

Nomogram for finding out quantity of hot gas required for
supplying given quantity of heat at different inlet temperatures of hot gas

Scale A - 1 cm = 0.5/1.0 million kcal/hr
B - 1 cm = 25/50x1000 nm³ / hr

DEOLALKAR CONSULTANTS

hot-gas temperature °C

400°C
300°C
270°C
200°C

x1000 nm³/hr

Nomo no.:	3-3-1 page 2
Title:	Nomogram for finding out quantity of hot gas required for supplying given quantity of heat at different inlet temperatures of hot gas
Useful for:	Finding out quantity of hot gas for supplying required quantity of heat and for designing system for it

Inputs:

1   heat required to be supplied in million kcal/hr

2   inlet temperature of hot gas ranging from 200- 400 $^\circ$ C

3   basis : ambient temperature = 35 $^\circ$ C
       sp. heat of gas = 0.3 kcal/nm^3
       gas cooled to - 90 $^\circ$ C

Output: quantity of hot gas in nm^3/hr

Scale:

A :   1 cm = 0.5/1.0 million kcal/hr
B :   1 cm = 25/50 × 1000 nm^3/hr

How to use:

From required quantity of heat on line A, draw a line through inlet temp. of hot gas on line 0-0, to meet line B and read quantity of hot gas required in nm^3/hr

Example:

heat required = 5 million kcal/hr (line A)
temp. of hot gas = 400 $^\circ$ C (line 0-0)
quantity of hot gas to be supplied = ~ 54000 nm^3/hr
                                    (line C)

Source: constructed

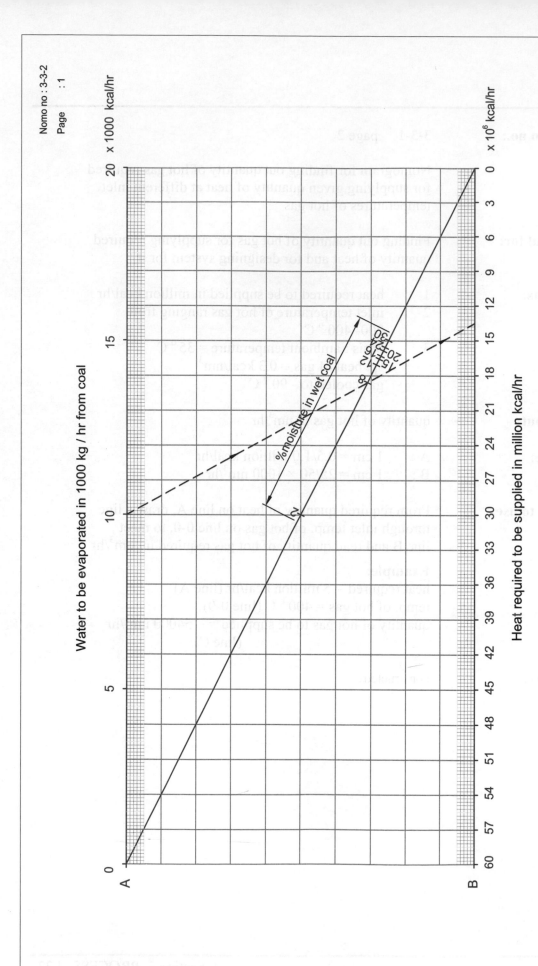

Nomo no : 3-3-2
Page   : 1

Water to be evaporated in 1000 kg / hr from coal

20 × 1000 kcal/hr

0   3   6   9   12   15   18   21   24   27   30   33   36   39   42   45   48   51   54   57   60   × 10⁶ kcal/hr

Heat required to be supplied in million kcal/hr

%moisture in wet coal

example : draw a line from 10000 kg/hr on line A through 8% on line 0-0 to meet line B. Read heat required 13.5 million kcal/hr

Nomogram for working out heat required for evaporating water in wet coal with different percentages of moisture

Scale : A: 1 cm = 1000 kg / hr
**   B: 1 cm = 3 million kcal / hr.**

DEOLALKAR CONSULTANTS

Nomo no.:	3-3-2 page 2
Title:	Nomogram for working out heat required to evaporate water in wet coal with different percentages of moisture
Useful for:	working out heat required to be supplied by hot gases for evaporating given quantities of water
Inputs:	1 quantities of water to be evaporated in kg/hr 2 % moisture in wet coal
Output:	heat required to be supplied in million kcal/hr
Scale:	A : 1 cm = 1000 kg/hr B : 1 cm = 3 million kcal/hr
How to use:	Draw a line from point indicating quantity of water to be evaporated on line A through point showing % moisture in wet coal on line 0-0 to meet line B and read heat required to be supplied in million kcal/hr (see Nomo no 3-3-4 or RS No 44 from 'Handbook for Designing Cement Plants' by S.P. Deolalkar for heat required in kcal to evaporate 1 kg of water at different moisture contents in feed)

Example:

quantity of water to be evaporated = 10000 kg/hr

(line A)

moisture in wet coal = 8 % (line 0-0)

heat required to be supplied = 13.5 million kcal/hr

(line B)

Source:	base data Otto Labahn nomogram constructed

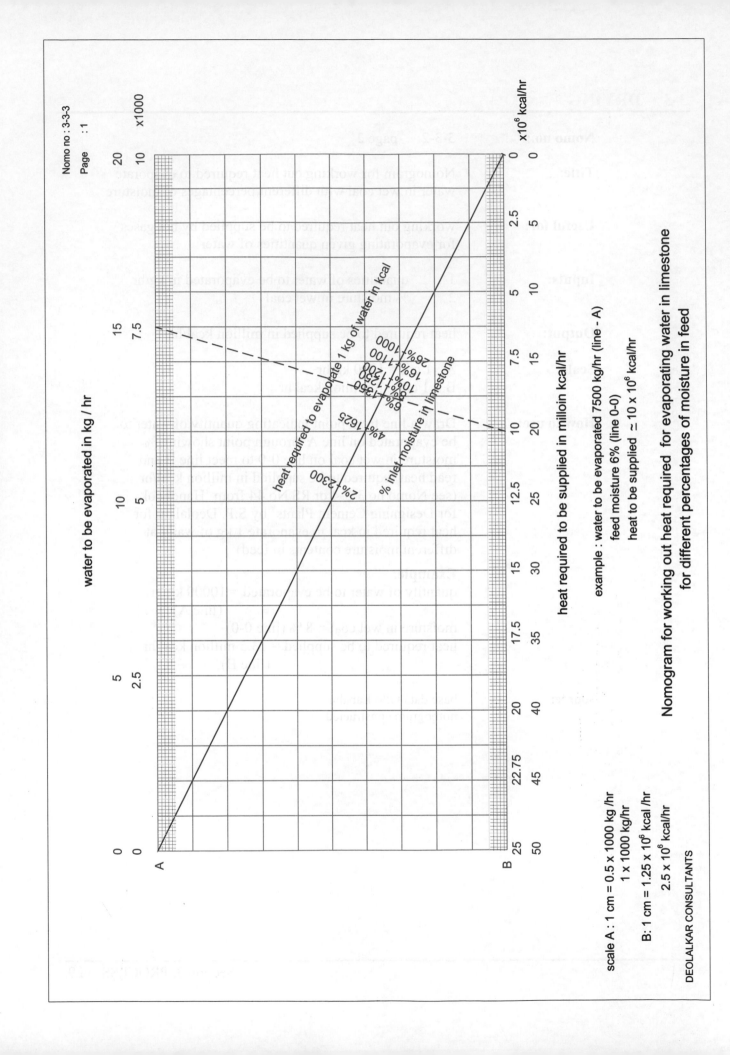

water to be evaporated in kg / hr

heat required to evaporate 1 kg of water in kcal

2% ~ 2300
4% ~ 1625
6% ~ 1350
8% ~ 1250
10% ~ 1200
16% ~ 1100
26% ~ 1000

% inlet moisture in limestone

heat required to be supplied in milloin kcal/hr

example : water to be evaporated 7500 kg/hr (line - A)
          feed moisture 6% (line 0-0)
          heat to be supplied ≈ 10 x 10⁶ kcal/hr

Nomogram for working out heat required for evaporating water in limestone
for different percentages of moisture in feed

scale A : 1 cm = 0.5 x 1000 kg /hr
               1 x 1000 kg/hr

       B: 1 cm = 1.25 x 10⁶ kcal /hr
               2.5 x 10⁶ kcal /hr

DEOLALKAR CONSULTANTS

Nomo no.:	3-3-3 page 2
Title:	Nomogram for working out heat required for water in limestone for different percentages of moisture in feed
Useful for:	calculating quantity of hot gases to supply the required quantity of heat
Inputs:	1 % moisture in feed 2 quantity of water to be evaporated in kgs / hr 3 heat required to drive 1 kg of water at different % of moisture in feed
Output:	quantity of heat required to be supplied in million kcal/hr
Scale:	A : 1 cm = 500/1000 kg/hr B : 1 cm = 1.25/2.5 million kcal/hr
How to use:	**Example:** Draw a line from quantity of water (7500 kg) to be evaporated on line A thru % moisture in feed (6% -1350) on line 0-0, to meet line B. Read heat required (~10.0 million kcal / hr) on it.
Source:	constructed

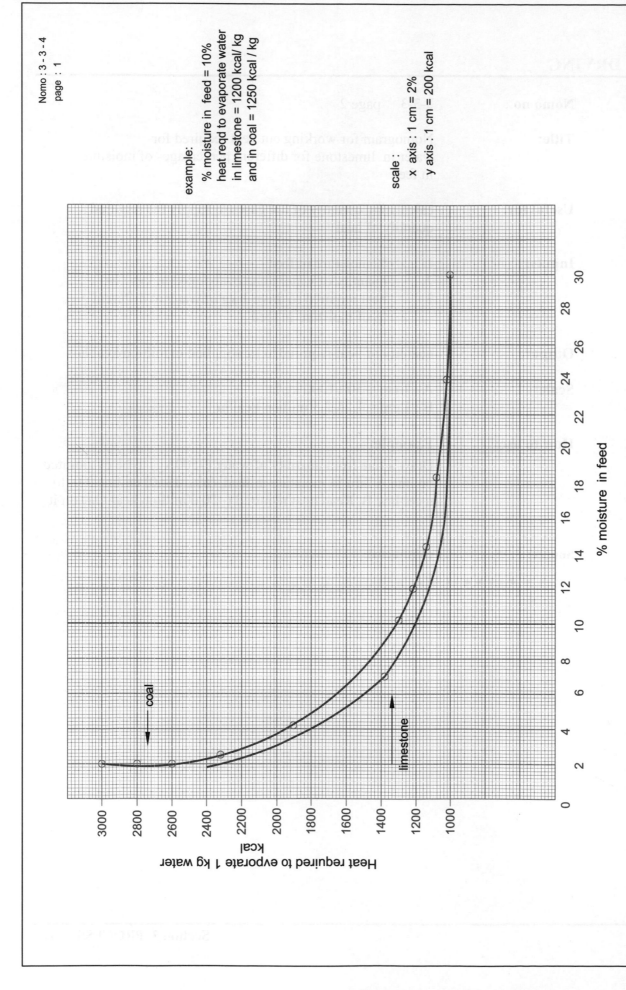

example:

% moisture in feed = 10%
heat reqd to evaporate water
in limestone = 1200 kcal/ kg
and in coal = 1250 kcal / kg

scale :
x axis : 1 cm = 2%
y axis : 1 cm = 200 kcal

Heat required to evporate 1 kg water
kcal

% moisture in feed

Graphs for heat required to evaporate 1 kg of water
in limestone & coal for different percentages of moisture in feed

coal

limestone

DEOLALKAR CONSULTANTS

Nomo no.:	3-3-4 page 2
Title:	Graphs for heat required to evaporate 1 kg of water in limestone and coal for different percentages of moisture in feed
Useful for:	arriving at heat required to be supplied for drying raw materials and coal
Inputs:	1 % moisture in feed
Output:	heat required in kcal/kg of water evaporated for limestone and coal
Scale:	x axis : 1 cm = 2 %
	y axis : 1 cm = 200 kcal/kg

see following table :

moisture %	heat required in kcal/kg of water evaporated	
	limestone	coal
2	2300	3000
4	1625	1750
6	1350	1500
8	1250	1350
10	**1200**	**1250**
12	1150	1200
14	1125	1150
16	1100	1125
18	1075	1100
20	1050	1075
22	1025	1030
24	1010	1020
26	1000	1010
28	1000	1000
30	1000	1000

Source:	Otto Labahn

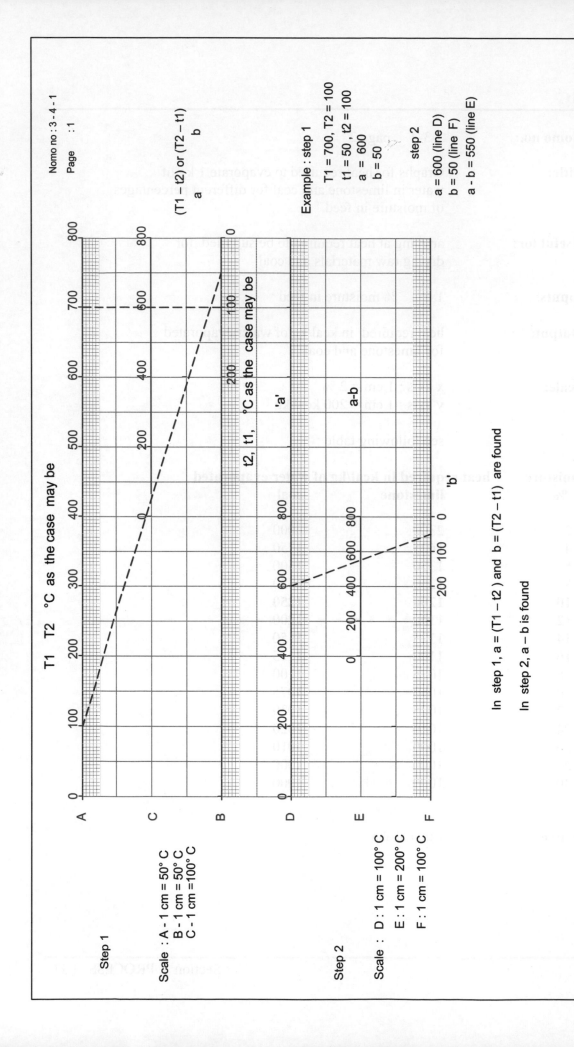

Nomogram for working out Log Mean Temperature Difference (L.M.T.D)

step 3

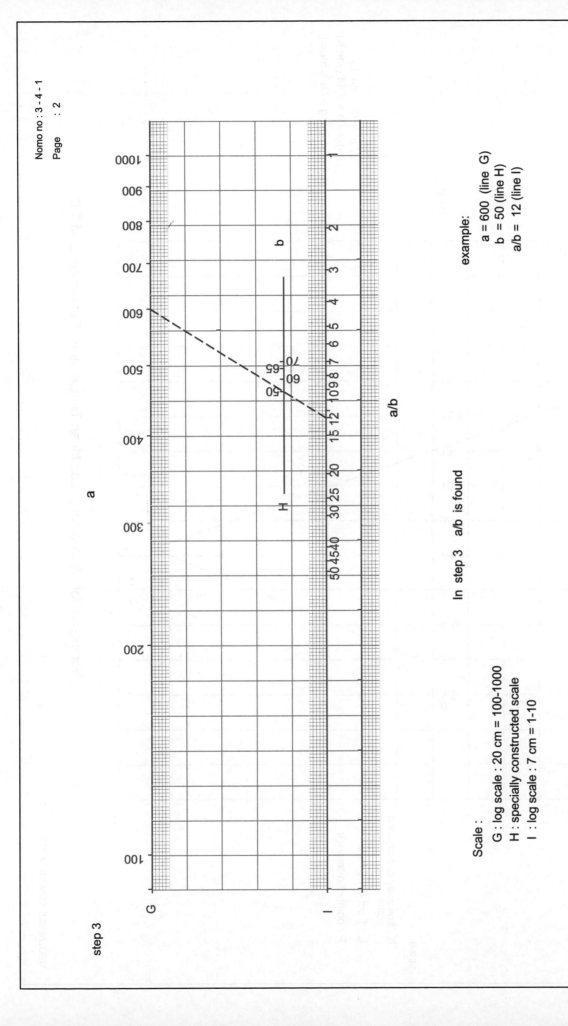

In step 3   a/b   is found

Scale :

G : log scale : 20 cm = 100-1000
H : specially constructed scale
I : log scale : 7 cm = 1-10

example:

a = 600 (line G)
b = 50 (line H)
a/b = 12 (line I)

Nomogram for working out Log Mean Temperature Difference (L.M.T.D.)

DEOLALKAR CONSULTANTS

Step 4

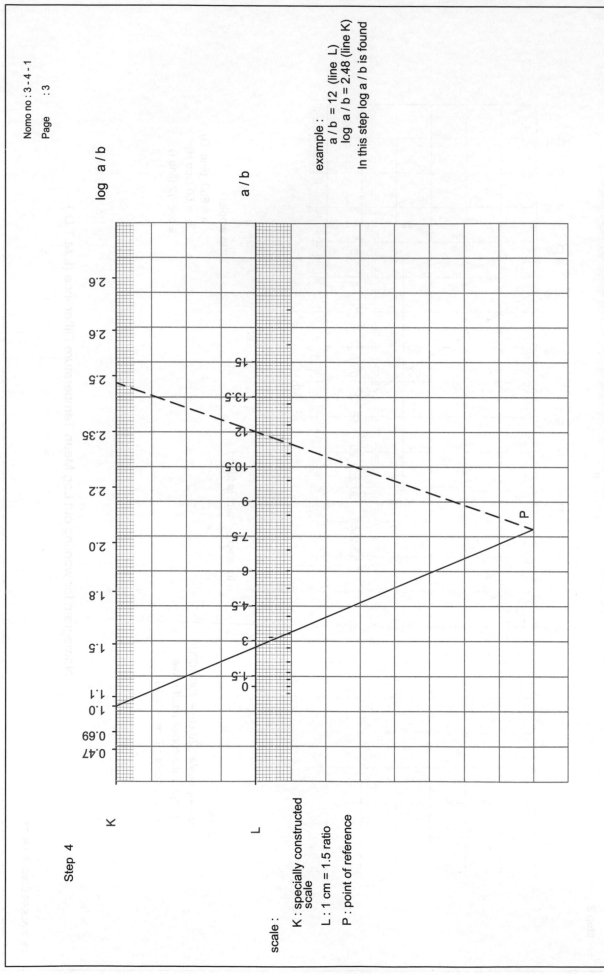

log a / b

a / b

K

L

scale :

K : specially constructed
scale

L : 1 cm = 1.5 ratio

P : point of reference

example :
a / b = 12 (line L)
log a / b = 2.48 (line K)
In this step log a / b is found

Nomogram for Working out Log Mean Temperature Difference L. M.T.D.

Step 5

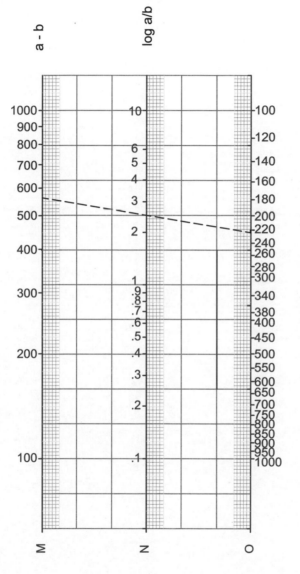

a - b

log a/b

L.M.T.D.

M

N

O

example : a – b = 600 (line M)

log a/b = 2.48 (line N)

l m t d = ~ 220 (line O)

Scale : M : log scale 10 cm = 100 -1000
         N : log scale 5 cm = 1 -10 & 0.1-1
         O : log scale 10 cm = 100 -1000

Nomogram for working out Log Mean Temperature Difference (L.M.T.D.)

DEOLALKAR CONSULTANTS

Nomo no.:   3-4-1   page 5

Title: Nomogram for working out Log Mean Temperature
                       Difference (LMTD)

Useful for: Sizing Rotary coolers and dryers where heat is
                       exchanged between material and gases

Inputs:

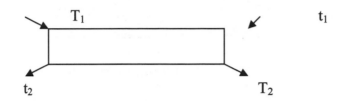

$$LMTD = (T_1 - t_2) - (T_2 - t_1) / \log_e (T_1 - t_2)/ (T_2 - t_1)$$

Let $(T_1 - t_2) = a$ and $(T_2 - t_1) = b$, then,
L.M.T.D. $= (a - b) / \log_e a/b$
where T_1 & T_2 are gas temperatures at inlet and outlet
t_1 and t_2 are material temperatures at inlet and outlet

Output: LMTD

                       Nomogram is constructed in 5 steps
                       1st step - find a and b
                       2nd step - find (a - b)
                       3rd step - find a/b
                       4th step - find $\log_e$ a/b
                       5th step - find LMTD

Scales: 1st step
                       A : 1 cm = 50 °C
                       B : 1 cm = 50 °C
                       C : 1 cm = 100 °C a or b
                       2nd step
                       D : 1 cm = 100 °C a
                       E : 1 cm = 200 °C a - b
                       F : 1 cm = 100 °C b
                       3rd step
                       G : log scale : 20 cm = 100-1000   a
                       H : specially constructed scale b
                       I : log scale : 7 cm = 1-10. 10-100   a / b

Nomo no.:	3-4-1 page 6
	4th step
	K : specially constructed scale for $\log_e$ a/b
	L : 1 cm = 1.5 a/b
	5th step
	M : log scale : 10 cm = 100-1000 a-b
	N : log scale : 5 cm = 0.1-1, 1-10 $\log_e$ a/b
	O : log scale : 10 cm = 100-1000 LMTD
How to use:	step 1
	from data received establish a and b
	read a or b on line C
	step 2
	from a on line D, draw a line to b on line F, read (a - b)
	on line E
	step 3
	from a on line G, draw a line thru b on line H to meet line
	I and read a/b
	step 4
	from reference point P, draw a line thru a/b on line L and
	read $\log_e$ a/b on line K
	step 5
	from point (a - b) on line M, draw a line thru $\log_e$ a / b on
	Line N and extend to meet line O and read LMTD
	Example:
	let T_1, T_2, t_1 & t_2 be 700, 100, 50 & 100 respectively, then
	a = 600 and b = 50, in step 1
	a - b = 550, in step 2
	a / b = 12, in step 3
	$\log_e$ a / b = 2.48 in step 4
	LMTD = ~ 220 in step 5
Source:	constructed

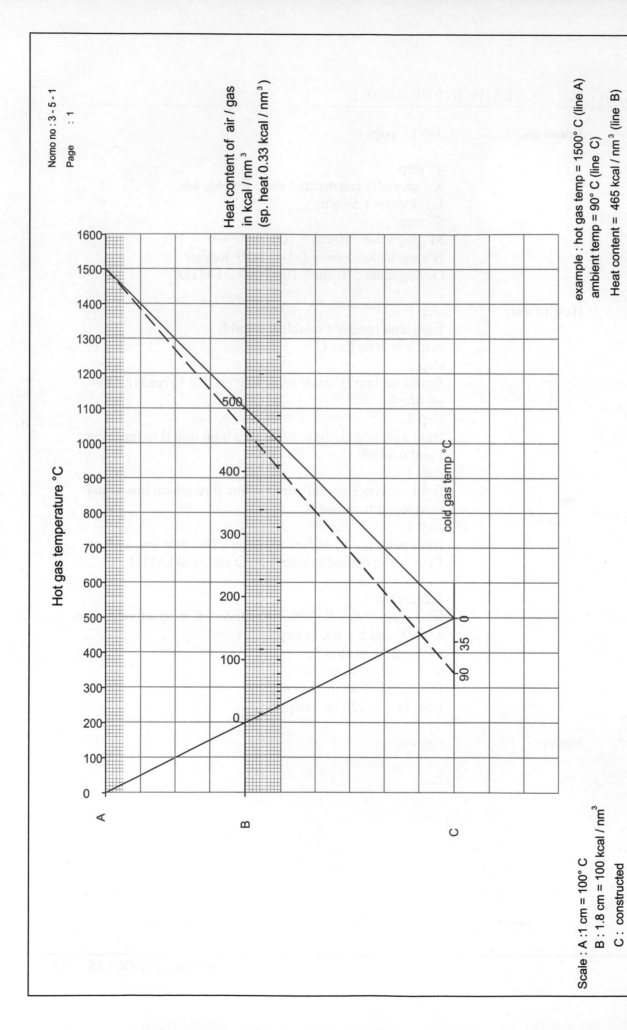

Heat content of air / gas
in kcal / nm³
(sp. heat 0.33 kcal / nm³)

Hot gas temperature °C

cold gas temp °C

example : hot gas temp = 1500° C (line A)
ambient temp = 90° C (line C)
Heat content = 465 kcal / nm³ (line B)

Nomogram for working out heat content of air/gas at different hot (inlet) & cold (outlet) temperatures

Scale : A :1 cm = 100° C
        B : 1.8 cm = 100 kcal / nm³
        C : constructed

DEOLALKAR CONSULTANTS

Nomo no.:	3-5-1 page 2
Title:	Nomogram for working out heat content of air/gas at different hot (inlet) and cold (outlet) temperatures
Useful for:	working out quantity of gas /air required for drying in mill etc. and for designing the system
Inputs:	1 hot gas temperature $^\circ$C 2 cold gas temperature $^\circ$C assumption – sp. heat of gas /air = 0.33 kcal/nm^3 on an average between 0-1500 $^\circ$C
Output:	useful heat content in kcal/ nm^3
Scale:	A : 1 cm = 100 $^\circ$C B : 1.8 cm = 100 kcal/ nm^3 C : constructed
How to use:	draw a line between inlet and outlet temperatures on lines A and C respectively and read heat content on line B in kcal/ nm^3

Example:
hot gas temp. = 1500 $^\circ$C line (A)
cold gas temp. = 90 $^\circ$C (line C)
heat content (useful) = ~ 465 $^\circ$C (line B)

Source:	constructed

Heat content of air/gas in kcal/nm³ (@ sp. heat 0.33 kcal / nm³ vide nomo 3-5-1)

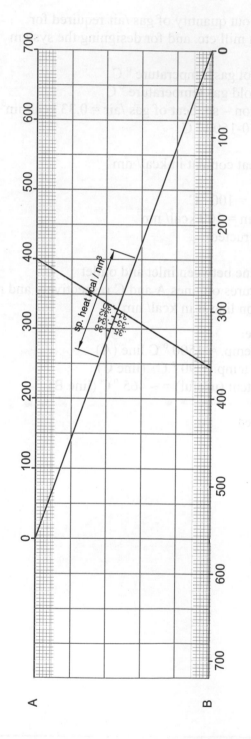

corrected heat content at actual sp. heat in kcal/nm³

sp. heat kcal / nm³

A

B

example : heat content 400 kcal /nm³ (line A)
sp. heat = 0.3 kcal / nm³ (line 0-0)
corrected heat content ≈ 364 kcal/ nm³ (line B)

Nomogram for correcting heat content of gas (vide nomo 3-5-1)
as per actual sp. heat of air / gas

Scale : A : 2 cm = 100 kcal /nm³
B : 2.5 cm =100 kcal/nm³

Nomo no.:	3-5-2 page 2
Title:	Correcting heat content of air / gas (vide nomo 3-5-1) as per actual sp. heat of air / gas
Useful for:	more accurate heat balance
Inputs:	1 heat content of air / gas as arrived at in nomo 3-5-1 at a sp. heat of 0.33 kcal/nm^3
	2 sp. heats of air / gas between 0.3 to 0.35 kcal/nm^3
Output:	corrected heat content of air/ gas
Scale:	A : 2 cm = 100 kcal/nm^3
	B : 2.5 cm = 100 kcal/nm^3
How to use:	determine heat content of air/gas vide nomo 3-5-1 at a sp. heat of 0.33
	Locate this point on line A and draw a line from it thru actual sp. heat on line 0-0 and extend it to line B and read corrected heat content in kcal/nm^3

Example:
heat content at 0.33 kcal /nm^3 = 400 (line A)
sp. heat = 0.30 kcal/ nm^3 (line 0-0)
corrected heat content $\simeq$ 364 kcal/nm^3

Source:	constructed

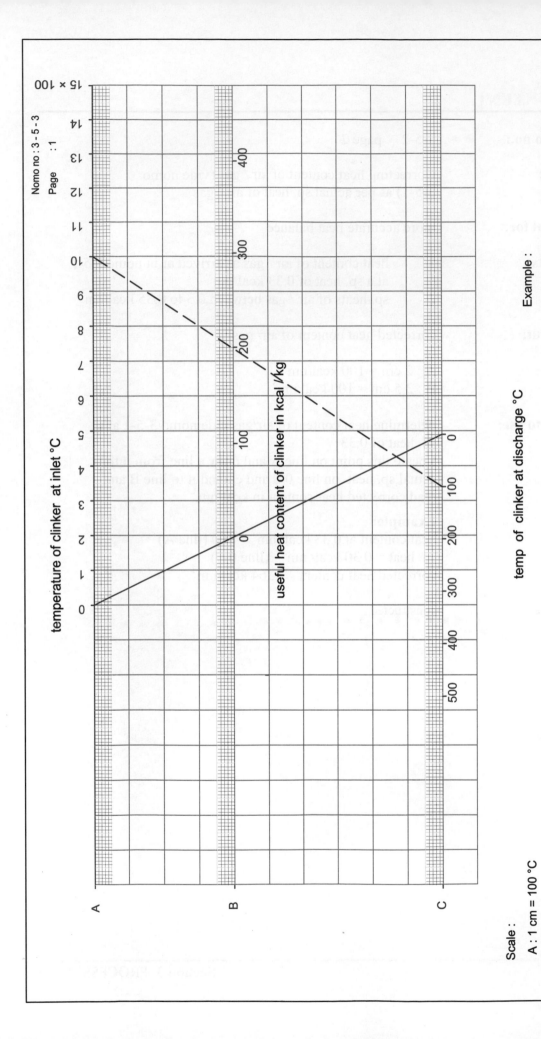

temperature of clinker at inlet °C

useful heat content of clinker in kcal /kg

temp of clinker at discharge °C

A

B

C

Nomo no : 3 - 5 - 3
Page   : 1

Example :

temp. of clinker = 1000 °C

temp. of clinker at discharge = 100 °C

heat content of clinker = ~ 200 kcal / kg

Nomogram for working out heat content of clinker in kcal / kg

Scale :

A : 1 cm = 100 °C

B : 2.7 cm =100 kcal / kg

C : constructed

Nomo no.:	3-5-3 page 2
Title:	Nomogram for working out heat content of clinker in kcal/kg
Useful for:	working out heat balance of kiln and cooler
Inputs:	1 temperature of hot clinker – 0-1500 ° C 2 sp. Heat of clinker 3 temperature of clinker at discharge end 0-500 ° C
Output:	Useful heat content of clinker in kcal/kg
Scale:	page 1 A : 1 cm = 100 ° C B : 2.7 cm = 100 kcal/kg C : constructed
How to use:	nomogram has been drawn for an average sp. heat of clinker 0.22 kcal/kg in the temperature range 0-1500 ° C draw a line from temp. of hot clinker on line A to temperature of cold clinker on line C and read heat content on line B.

Example:

temp. of clinker at inlet = 1000 ° C		(line A)
temp. of clinker at discharge = 100 ° C		(line C)
heat content	= ~ 200 kcal /kg	(line B)

Source:	constructed

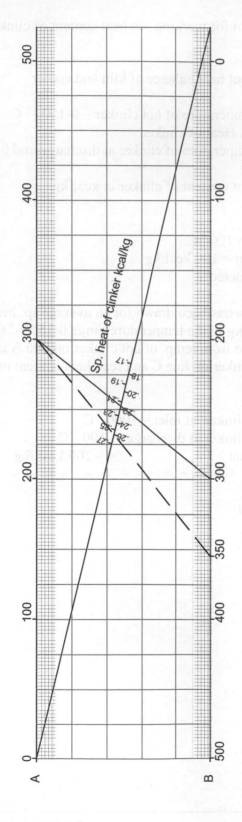

Heat content of clinker in kcal/kg (vide nomo 3 - 5 - 3)

Corrected heat content of clinker kcal/kg

Sp. heat of clinker kcal/kg

A

B

Scale : A : 1 cm = 25 kcal / kg
        B : 1 cm = 25 kcal / kg

example : heat content arrived - 300 kcal / kg (at specific heat 0.22 kcal / kg) (line A)
actual sp/ heat = 0.26 (line 0-0) (see Table in nomo 3-5-3)
corrected heat content = ≃ 355 kcal/kg (line B)

Nomogram for correction for heat content of clinker worked out in nomo (3-5-3)

DEOLALKAR CONSULTANTS

Nomo no.:	3-5-4 page 2
Title:	correction for heat content of clinker vide nomo 3-5-3
Note:	heat content of clinker in nomo 3-5-3 was based on average sp. heat of clinker of 0.22 kcal/kg. It actually varies between 0.17 to 0.27 kcal/kg in the temperature range 0-1500 °C
Useful for:	more accurate workouts
Inputs:	1 heat content of clinker at sp. heat of 0.22 2 actual sp. heats of clinker at different temperatures (see table on page 3)
Output:	corrected sp. heat of clinker in kcal/kg
Scale:	A : 1 cm = 25 kcal/kg B : 1 cm = 25 kcal/kg
How to use:	Ascertain heat content of clinker at a sp. heat of 0.22 vide nomo 3-5-3. Locate this value on line A. Find actual sp. heat of clinker from graph on page 2 of nomo 3-5-3. Draw a line from point on line A thru actual sp. heat on line 0-0 and extend to meet line B and read corrected heat content

Example:
heat content vide nomo 3-5-3 = 300 kcal/kg   (line A)
actual sp. heat of clinker = 0.26   (line 0-0)
corrected heat content = ~355 kcal/kg   (line B)

Source:	constructed

Nomo no.:　　　　3-5-4　page 3

Table 1

Table of SP. Heats of Air and Clinker at different temperatures

Temperature	sp. heat of air	sp. Heat of clinker
°C	kcal/nm^3	kcal/kg
0	0.31	0.173
100	0.312	0.188
200	0.313	0.2
300	0.315	0.206
400	0.318	0.213
500	0.321	0.218
600	0.324	0.224
700	0.328	0.228
800	0.33	0.231
900	0.334	0.234
1000	0.337	0.237
1100	0.34	0.242
1200	0.343	0.248
1300	0.345	0.253
1400	0.347	0.261
1500	0.35	0.265

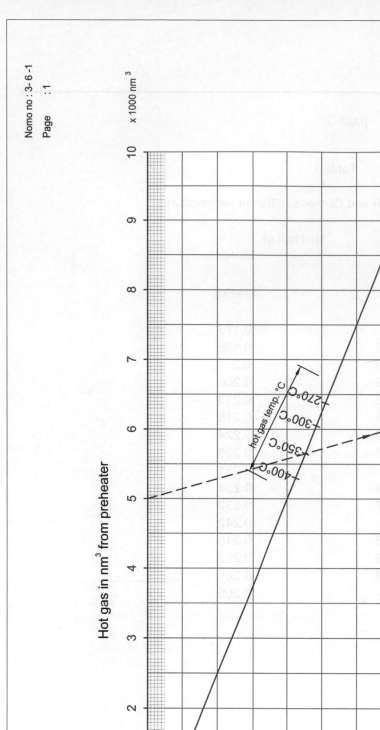

Hot gas in nm³ from preheater

x 1000 nm³

x 1000 nm³

Resultant gas including diluting air cooled to 240° C

Example :

hot gas = 5000 nm³ (line A)

hot gas temp. = 350° C (line 0-0)

resultant exit gas = 7700 nm³ at 240°C (line B)

Nomogram for working out resultant gas volume after addition of dilution air
to cool gases to 240° C, a suitable for a glass bag filter

Scale : A : 1 cm = 500 nm³
B : 1 cm = 1000 nm³

Nomo no.:	3-6-1 page 2
Title:	Nomogram for working out resultant gas volume after addition of diluting air to cool gases to 240 ° C suitable for glass bag filter
Useful for:	finding out readily volume of cold gases and for designing bag filter system
Inputs:	1 quantity of hot gas in nm^3 at different temperatures (400, 350, 270 ° C)
	2 temperature to which to be cooled- 240 ° C
	3 assumed sp. heat of gas / air 0.3 kcal/ nm^3
	4 ambient temperature 35 ° C
Output:	Resultant gas in nm^3 at 240 ° C
Scale:	A : 1 cm = 500 nm^3
	B : 1 cm = 1000 nm^3
How to use:	Draw a line from quantity of hot gas on line A through hot gas temperature on line 0-0 to meet line B and read resultant quantity of gas at 240 ° C

Example:
quantity of hot gas = 5000 nm^3 (line A)
temperature of hot gas = 350 ° C (line 0-0)
resultant gas at 240 ° C = ~ 7700 nm^3 (line B)
therefore quantity of diluting air = 2700 nm^3

Source:	constructed

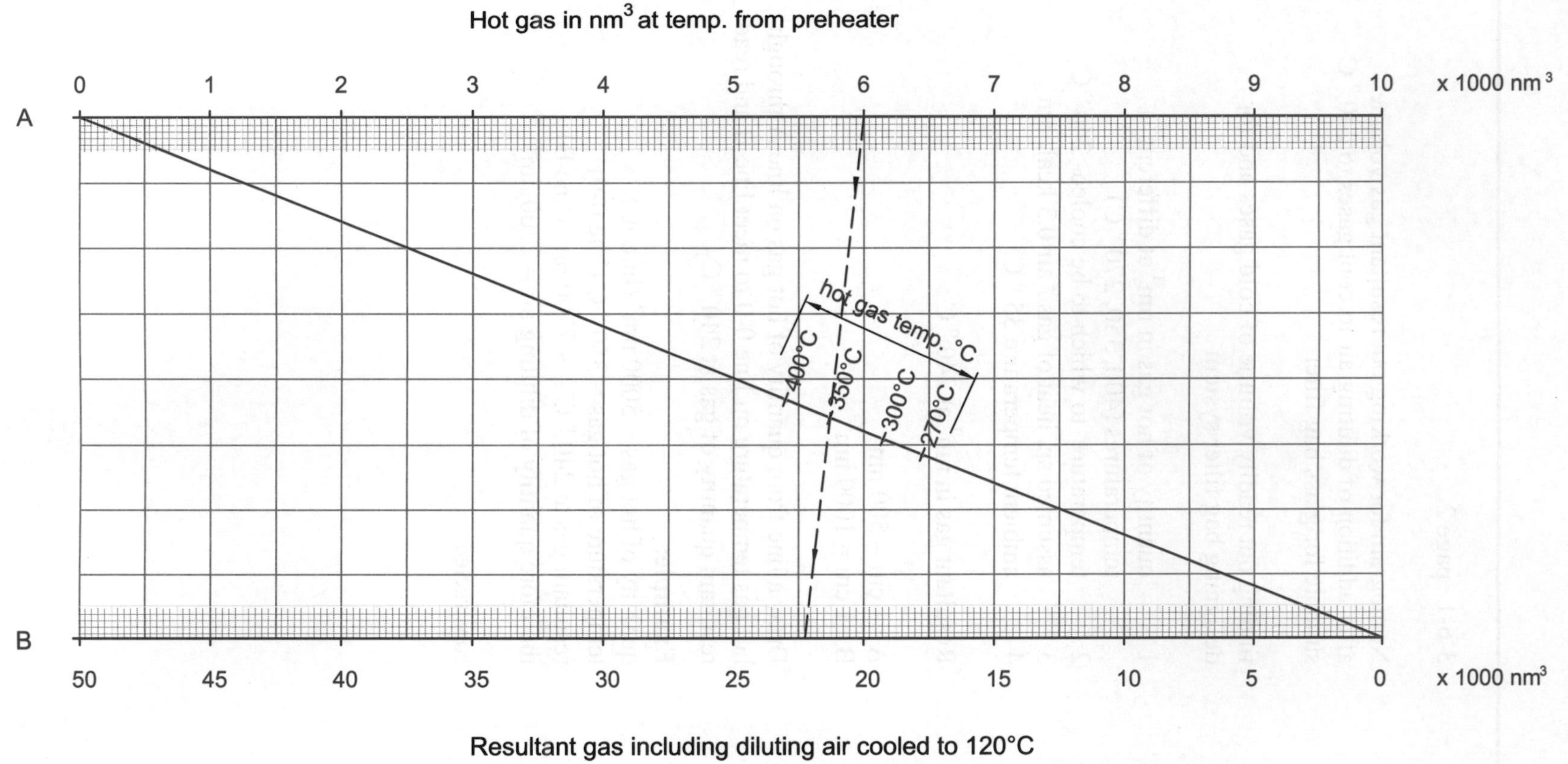

Hot gas in nm³ at temp. from preheater

x 1000 nm³

x 1000 nm³

Resultant gas including diluting air cooled to 120°C

hot gas temp. °C

270°C
300°C
350°C
400°C

Example :

hot gas = 6000 nm³   (line A)
temp. = 350° C (line 0-0)
resultant gas at 120 ° C = 22200 nm³ (line B)

Nomogram for working out quantity of resultant gas after addition of diluting air to cool gases to 120 ° C,
suitable for a bag filter with polyester bags

Scale : A: 1cm = 500 nm³
B: 1cm = 2500 nm³

DEOLALKAR CONSULTANTS

Nomo no.:	3-6-2 page 2
Title:	Nomogram for working out quantity of resultant gas after addition of diluting air to cool gases to 120 $^\circ$C suitable for bag filter with polyester bags
Useful for:	finding readily quantity of gas cooled to 120 $^\circ$C and for designing bag filter
Inputs:	1 hot gas in nm^3 at different temperatures (400, 350, 300, 270 $^\circ$C)
	2 temperature to which it is to be cooled (120 $^\circ$C)
	3 assumptions : sp. heat of gas and air 0.32 kcal/ nm^3 ambient temperature 35 $^\circ$C
Output:	resultant quantity of gas in nm^3 at 120 $^\circ$C
Scale:	A : 1 cm = 500 nm^3
	B : 1 cm = 2500 nm^3
How to use:	Draw a line from quantity of hot gas on line A through temperature of hot gas on line 0-0 to meet line B. Read quantity of resultant gas at 120 $^\circ$C

Example:
Quantity of hot gas = 6000 nm^3 (line A)
Temperature of hot gas = 350 $^\circ$C (line 0-0)
Quantity of resultant gas at 120 $^\circ$C = ~ 22250 nm^3

Source:	constructed

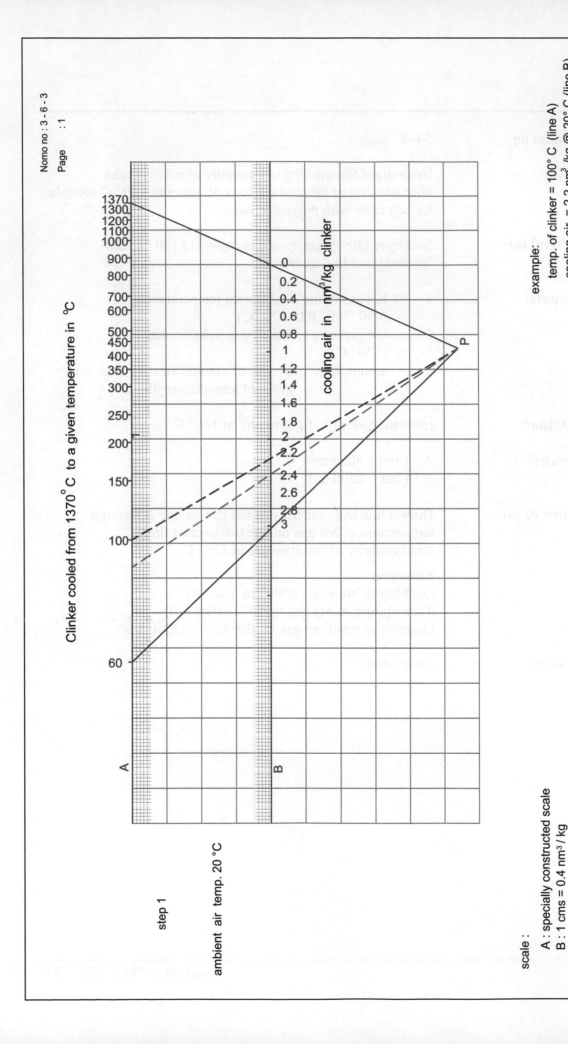

step 1

Clinker cooled from 1370° C to a given temperature in °C

1370
1300
1200
1100
1000
900
800
700
600
500
450
400
350
300
250
200
150
100
60

A

ambient air temp. 20 °C

cooling air in nm³/kg clinker

0
0.2
0.4
0.6
0.8
1
1.2
1.4
1.6
1.8
2
2.2
2.4
2.6
2.8
3

B

P

example:
temp. of clinker = 100° C (line A)
cooling air = 2.2 nm³/kg @ 20° C (line B)

Nomogram for cooling air at different temperatures in nm³ / kg of clinker for
cooling it from 1370 ° C to a given temp. in conventional coolers

scale :
A : specially constructed scale
B : 1 cms = 0.4 nm³ / kg
P : point of reference

DEOLALKAR CONSULTANTS

Nomo no.:	3-6-3 page 2
Title:	Nomogram for cooling air at different temperatures in nm^3/kg clinker for cooling it from 1370 $^\circ$C to a given temp. in conventional coolers
Useful for:	heat balance of coolers
Inputs:	1 temperature of clinker $^\circ$C 2 discharge temp. of clinker $- 60 ^\circ$C 3 ambient temperature 20 $^\circ$C
Output:	cooling air required in nm^3/ kg clinker
Scale:	page 1 A : specially constructed scale for clinker temp. at inlet between 1370 and 60 $^\circ$C B : 1 cm = 0.4 nm^3/ kg P : point of reference page 3 C : 1 cm = 0.20 nm^3/ kg D : 1 cm = 0.20 nm^3/ kg E : 1 cm = 0.20 nm^3/ kg
How to use:	**Example:** draw a line from temp. of cooled clinker (100 $^\circ$C) on line A to point of reference P. It intersects line B. point of intersection , (2.2) is cooling air required to cool clinker from 1370 to 100 $^\circ$ C, in nm^3/ kg when ambient air temp. is 20 $^\circ$ C if ambient temp. is 40 $^\circ$ C, draw a line from 2.2 on line C on page 3 thru 40 $^\circ$ C on line 0-0 and extend to meet line D. read cooling air required as 2.4 nm^3/ kg. add another 10 % when cooling air from line B is between 2.25 and 3 for instance, for cooling clinker to 90 $^\circ$ C, cooling air would be 2.4 . for ambient temp. of 40 $^\circ$ C, this Would be ~ 2.6 nm^3/ kg. Add another 10 % and read cooling air as ~ 2.90 on line E
Source:	CPAG Manual Nomogram constructed

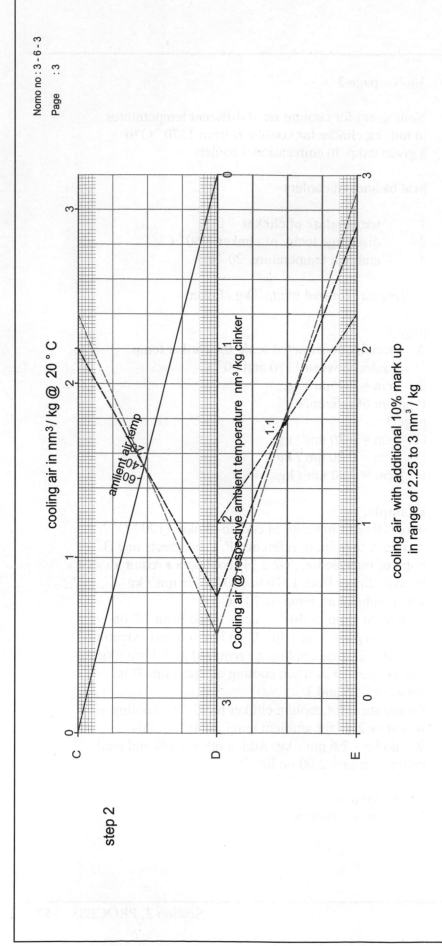

cooling air in nm³/ kg @ 20 ° C

step 2

Scale : C :1 cm = 0.2nm³ / kg
D : 1 cm = 0.2 nm³ / kg
E : 1 cm = 0.2 nm³ / kg

example: cooling air - say 2.2 nm³/ kg @ 20° C from step (line C)
ambient temperature 40° C, increase cooling air ≈ 2.4 nm³/ kg (line D)
add another 10%   2.65 nm³ (line E)
see text for further explanation
follow green lines when clnker is to be cooled to 90°C

Nomogram for cool ing air at different temperatures in nm³/ kg clinker for cooling it
from 1370 ° C to a given temp. in conventional coolers

Nomo no.:	3-6-3 page 4
Title:	Nomogram for cooling air at different temperatures in nm³/kg clinker for cooling it from 1370 °C to a given temp. in conventional coolers
Useful for:	heat balance of coolers
Inputs:	1 temperature of clinker °C 2 discharge temp. of clinker – 60° C 3 ambient temperature 20° C
Output:	cooling air required in nm³/ kg clinker

Scale:

page 1
A : specially constructed scale for clinker temp.
    at inlet between 1370 and 60 ° C
B : 1 cm = 0.4 nm³/ kg
P : point of reference
page 3
C : 1 cm = 0.20 nm³/ kg
D : 1 cm = 0.20 nm³/ kg
E : 1 cm = 0.20 nm³/ kg

How to use:

Example:
draw a line from temp. of cooled clinker (100 ° C)
on line A to point of reference P. It intersects line B.
point of intersection, (2.2) is cooling air required
to cool clinker from 1370 to 100 ° C, in nm³/ kg
when ambient air temp. is 20 ° C
if ambient temp. is 40 ° C, draw a line from 2.2 on
line C on page 3 thru 40 ° C on line 0-0 and extend to
meet line D. read cooling air required as 2.4 nm³/ kg.
add another 10 % when cooling air from line B is
between 2.25 and 3
for instance, for cooling clinker to 90 ° C, cooling air
would be 2.4 . for ambient temp. of 40 ° C, this
Would be ~ 2.6 nm³/ kg. Add another 10 % and read
cooling air as ~ 2.90 on line E

Source:

CPAG Manual
Nomogram constructed

Nomo no.:	3-5 page 4
Title:	Nomogram for cooling air at different temperatures in mm³/kg clinker for cooling it from 1350°C to a given temp. in conventional coolers
Useful for:	heat balance of coolers
Input:	1. temperature of clinker °C 2. discharge temp. of clinker – 60°C 3. ambient temperature 20°C
Output:	cooling air required in mm³/kg clinker
Scale:	page 1 A. specially constructed scale for clinker temp. at inlet between 1370 and 60°C B. 1 cm = 0.4 mm³/kg P: point of reference page 3 C: 1 cm = 0.20 mm³/kg D: 1 cm = 0.20 mm³/kg E: 1 cm = 0.20 mm³/kg
How to use:	**Example:** draw a line from temp. of cooled clinker (100°C) on line A to point of reference P; it intersects line B, point of intersection, (2.2) is cooling air required to cool clinker from 1370 to 100°C in mm³/kg, when ambient air temp. is 20°C. if ambient temp. is 30°C, draw a line from 2.2 pm. line B on page 3 thru 40°C on line C-D and extend to meet line D; read cooling air required as 2.4 mm³/kg add another 10% when cooling air from line B is between 2.2 and 2. for instance, for cooling clinker to 90°C, cooling air would be 2.4. for ambient temp. of 40°C, this would be = 2.6 mm³/kg. Add another 10% and read cooling air as = 2.90 on line E.
Source:	CPMC Manual Nomogram construction

Section 4 MACHINERY

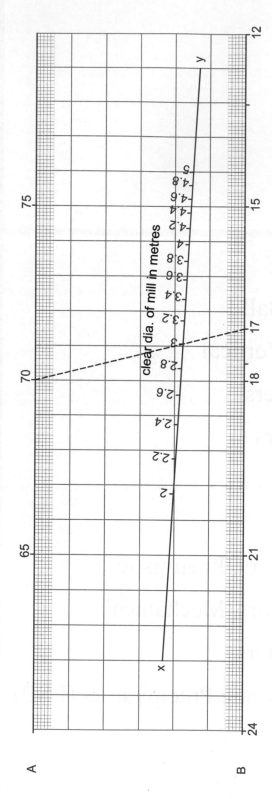

% critical speed

clear dia. of mill in metres

Actual mill speed r. p. m

Example: % critical speed = 70 (line A)
clear dia of mill = 3m (line xy)
speed of mil = ~ 17 r. p. m (line B)

Nomogram for working out speeds of ball mills in r. p. m
for different values of critical speeds and
clear diameters of mills

Scale : A :1 cm = 1 %
B :1 cm = 0.6 r. p. m

DEOLALKAR CONSULTANTS

Nomo no.:	4-1-1 page 2
Title:	Nomogram for speeds of ball mills in r.p.m. for different values of critical speeds and clear diameters of mills
Useful for:	sizing gear boxes
Inputs:	1 effective diameters of mills in metres 2 % of critical speeds
Output:	speeds of mills in r.p.m.
Scale:	A : 1 cm = 1 % B : 1 cm = 0.6 r.p.m.
How to use:	draw a line from desired value of critical speed on line A thru effective dia. of mill on line x-y and extend it to meet line B and read speed of mill in r.p.m.

Example:
speed of mill = 70 % of critical speed (line A)
effective dia. of mill = 3 m (line x-y)
speed of mill in r.p.m. = ~ 17 (line B)

Source:	constructed

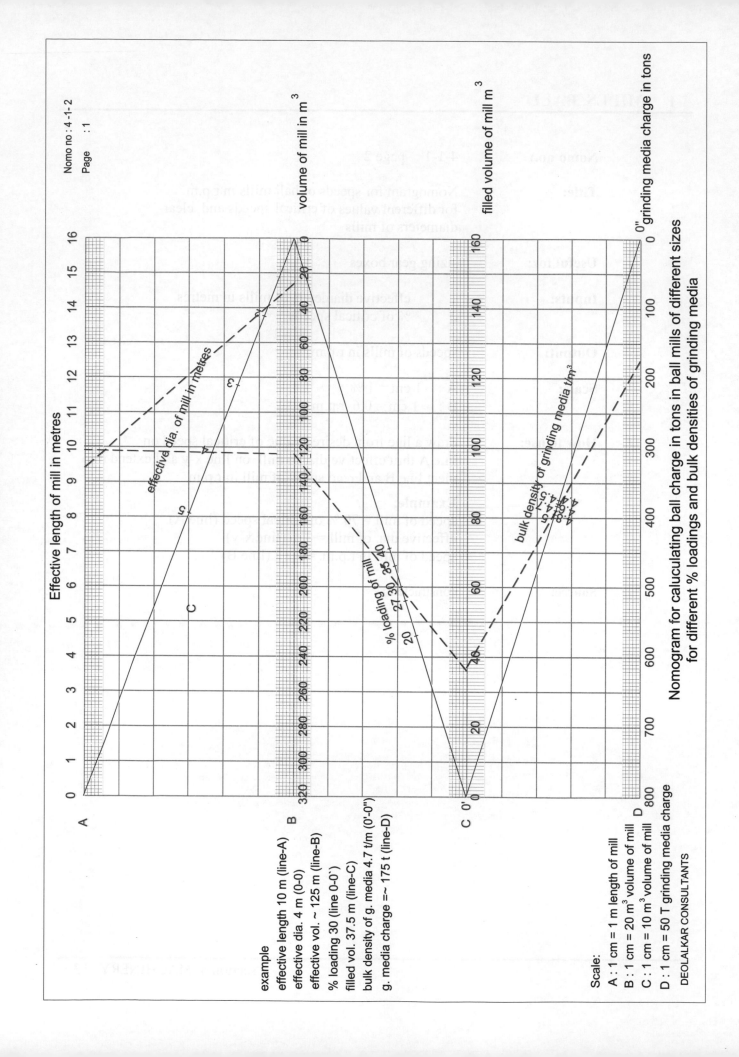

Effective length of mill in metres

volume of mill in m³

filled volume of mill m³

0" grinding media charge in tons

effective dia. of mill in metres

bulk density of grinding media t/m³

% loading of mill

Nomogram for caluculating ball charge in tons in ball mills of different sizes
for different % loadings and bulk densities of grinding media

example
effective length 10 m (line-A)
effective dia. 4 m (0-0)
effective vol. ~ 125 m (line-B)
% loading 30 (line 0-0')
filled vol. 37.5 m (line-C)
bulk density of g. media 4.7 t/m (0'-0")
g. media charge =~ 175 t (line-D)

Scale:
A : 1 cm = 1 m length of mill
B : 1 cm = 20 m³ volume of mill
C : 1 cm = 10 m³ volume of mill
D : 1 cm = 50 T grinding media charge
DEOLALKAR CONSULTANTS

Nomo no.:	4-1-2 page 2
Title:	Nomogram for calculating ball charge in tons in ball mills of different sizes for different % loadings and bulk densities of grinding media
Useful for:	arriving at total tonnage of grinding media compartment wise and as a whole in mills of different sizes and % loadings
Inputs :	1 mill size : length × diameter 2 % loading 3 bulk density of grinding media in t/m^3
Output:	quantity of grinding media in tons
Scale:	A : 1 cm = 1 m length of mill B : 1 cm = 20 m^3 volume of mill C : 1 cm = 10 m^3 volume of mill D : 1 cm = 50 ton weight of grinding media
How to use:	From given length of mill on line A draw a line thru effective dia. of mill on line 0-0 and extend to meet line B and read effective volume of mill. From this point draw a line thru % loading on line 0-0' to and extend to meet line C and read filled volume of mill. From this point draw a line thru bulk density of grinding media on line 0'-0″ and extend to line D and read load of grinding media in tons

Example:
effective length of mill = 10 m (line A)
effective dia. of mill = 4 m (line 0-0)
effective volume of mill = 125 m^3 (line B)
% loading in mill = 30 (line 0-0')
filled volume = 37.5 m^3 (line C)
bulk density of grinding media = 4.7 t/ m^3 (line 0'-0″)
weight of grinding media = ~ 175 t (line D)

Source:	constructed

Power at mill shaft in kw/metre

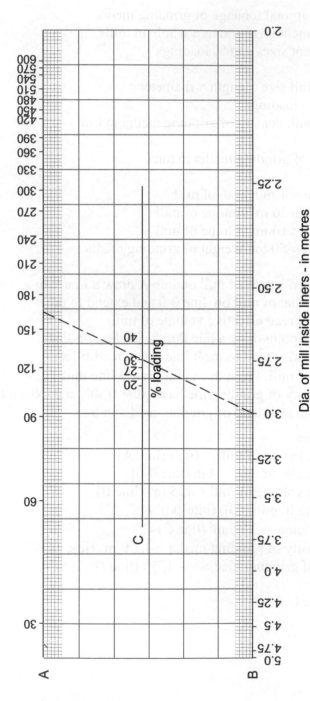

Scale : A : specially constructed scale for power in kw/metre

B : specially constructed scale for clear dia of mill in metres

C : % loading

example :

dia. of mill = 3 m (line B)

% loading = 30% (line C)

power in kw/metre = ~ 170

Nomogram for working out power at shaft in kw/metre for ball mills
of different diameters at different loads and at 75% critical speed

Nomo no.:	4-1-3 page 2
Title:	nomogram for working out power at shaft in kw / m for ball mills of different diameters at different loads at 75 % critical speed
Useful for:	quick reckoner of power required by a mill knowing its size and % loading
Inputs:	1 dia. of mill inside liners 2 speed assumed 75 % critical speed 3 % loading between 20 to 40 %
Output:	power required in kw /m at the mill for different loads
Scale:	A : specially constructed scale for power in kw/m B : specially constructed scale for clear diameter of mill in metres C : % loading
How to use:	draw a line from dia. of mill on line B, thru % loading on line C and extend to line A and read power drawn in kw / m.

Example:
clear dia. of mill = 3 m (line B)
% loading = 30 % (line C)
power at mill shaft in kw / m = ~ 170 (line A)

Source:	constructed

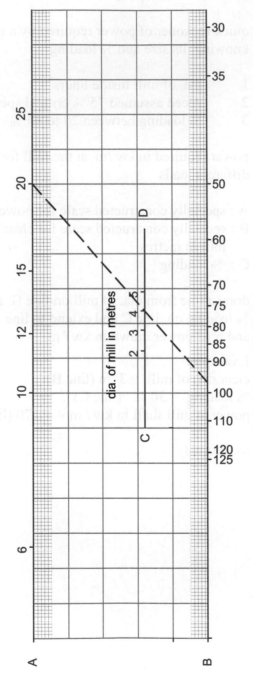

Feed size 80% passing in mm

dia. of mill in metres

dia. of largest ball in mm

A

B

example : size of feed 80% passing = 20 mm (line A)
dia of mill = 4 m (line CD)
dia of largest ball ≈ 100 mm (line B)

Nomogram for finding out largest diameter of ball in a ball mill
according to size of mill and size of feed

Scale : A : log scale 20 cm = 1 - 10 , 10 - 100 etc
         B : log scale 20 cm = 1 - 10

Nomo no.:	4-1-4 page 2
Title:	Nomogram for finding largest diameter of ball in a ball mill according to size of mill and size of feed
Useful for:	quick reckoning of largest dia. of ball. Pattern of charge of grinding media can then be worked out
Inputs:	1 size of feed in terms of 80 % passing in mm 2 diameter of mill in mm
Output:	diameter of largest ball in mm
Scale:	A : log scale – 20 cm = 1-10, 10-100 etc B : log scale – 20 cm= 1-10
How to use:	draw a line from size of feed on line A thru dia. of mill on line C to meet line B. Read dia. of ball in mm

Example:
feed size – 80 % passing = 20 mm (line A)
dia of mill (nominal) = 4 m (line C)
dia. of largest ball = ~ 100 mm
Note : not very accurate because dia. refers to nominal dia.
where as it should be clear dia.
Size of ball arrived at by formula rounded off.
In actual practice nearest available commercial size
would be used

Source:	formula for ball size from Duda nomogram constructed

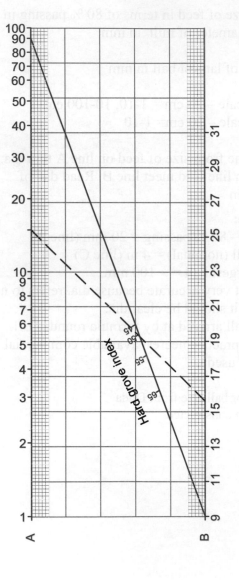

Scale : A : log scale 7 cm = 1-10, 10-100
B : 1 cm = 2 kwh / t

example :
fineness 15% residue on 90 microns (line A)
HGI. 50. (9-90)
sp. power ≃ 15.6 kwh / t (line B)

Nomogram for sp. power for a ball mill when grinding
materials of different hardness (HGI) to different finenesses

Nomo no.:	4-1-5 page 2
Title:	Nomogram for sp. power for a ball mill grinding materials of different hardness (HGI) to different finenesses
Useful for:	sizing ball mills
Inputs:	1 fineness of product - % residue on 90 microns 2 Hard Grove Indices (HGI)
Output:	sp. power in kwh / ton
Scale:	A : log scale – 7 cm = 1-10, 10-100 product fineness B : 1 cm = 2 kwh / ton
How to use:	**Example:** draw a line from given product fineness on line A (15), thru HGI on line 9-90 (50), and extend to line B and read sp. power on it (15.6 kwh / ton)
Source:	base data from FLS Manual nomogram constructed

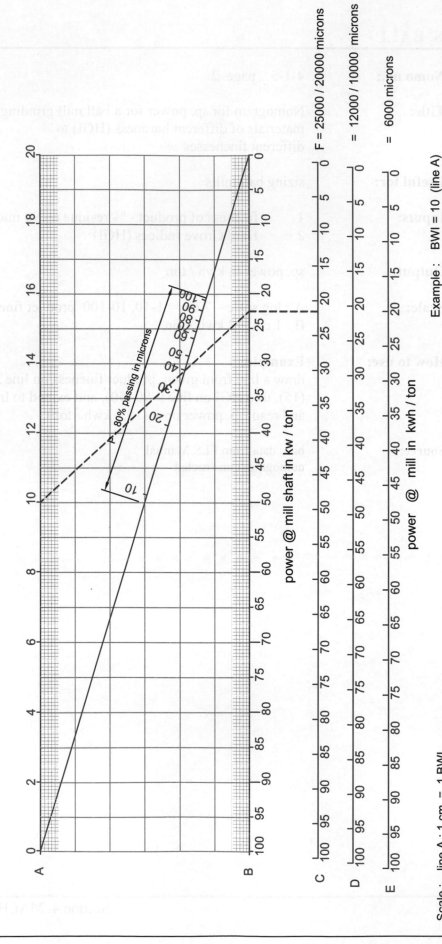

Nomo no : 4 - 1 - 6
Page : 1

BWI - in kwh / short ton

80% passing in microns

power @ mill shaft in kw / ton

power @ mill in kwh / ton

F = 25000 / 20000 microns

= 12000 / 10000 microns

= 6000 microns

Example :   BWI = 10 (line A)
           P = 40 micron (0 - 0)
           power at shaft = 22.5 kwh /t
           feed size = 20,000 microns
           sp. power in kwh / t = 21.5 kwh /t (line C)

Nomogram for working out sp. power of a ball mill knowing
Bond's Index, feed size and product fineness
of material to be ground

Scale :   line A : 1 cm = 1 BWI
          line B, C, D, & E : 1 cm = 5 kwh/t

DEOLALKAR CONSULTANTS

Nomo no.:	4-1-6 page 2
Title:	Nomogram for working out sp. power of a ball mill knowing BWI, feed size and product fineness of material to be ground.
Useful for:	sizing ball mills
Inputs;	1 material to be ground
	2 its Bond's Index (B.W.I.)
	3 product fineness P – 80% passing in microns
	4 feed size F - 80% passing in microns
Output:	sp. power for mill at shaft in kwh/ton
Note:	sp. power for ball mill for dry grinding

$$= 1.1 \times 1.3 \times BWI \times 10/P^{0.5} \times (1/P^{0.5} - 1/F^{0.5}) \text{ kwh/t}$$
$$= 14.3 \times a \times b \text{ where } a = BWI /P^{0.5} \text{ and } b = (1/P^{0.5} - 1/F^{0.5})$$

Nomogram in two steps :
1st Step = 14.3 × a
2nd Step = b
P ranges between 10 to 100 microns
F ranges between 25000 and 6000 microns
Reduction in sp. power when feed size is reduced
from 25000 microns to 6000 microns is small –
2-3 % and hence can be ignored in most cases for
sizing ball mills.

Scale:	1st step
	A : 1 cm = 1 BWI
	B : 1 cm = 5 kwh/t
	2nd step
	C, D, E : 1 cm = 5 kwh/t
How to use:	draw a line from given BWI on line A thru given product fineness (P) on line 0-0 and extend to line B and read sp. power in kwh/ton to allow for feed size, extend line vertically thru lines C, D or E corresponding to size of feed and read sp. power in kwh/t on it.

4-1-6   page 3

Example:
BWI = 10 (line A)
P = 40 microns (line 0-0)
Power at shaft = ~22.5 kwh/t (line B)
Feed size = 20000 microns
sp. power = 21.5 kwh/t (line C)
On line D
As mentioned above for practical purposes, mill could be sized
for sp. power of 22.5 kwh/t ignoring feed size

Source: constructed

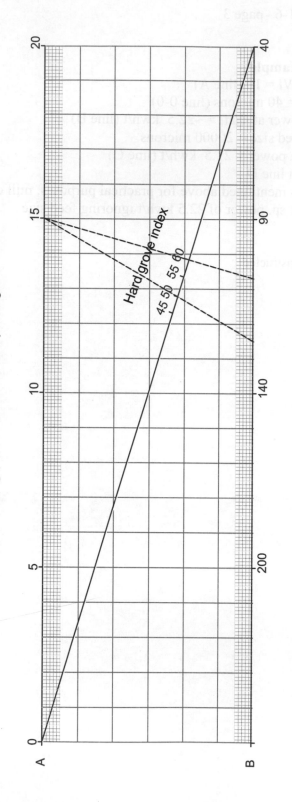

Thruput in tph @ product fineness 70% passing 200 mesh

Mill power in KW

Example :   thruput = 15 tph (line A)

              HGI = 50 (line 0-40)

              power for mill = 125 kw (line B)

              HGI = 60 (line 0-40)

              power for mill = 107 kw

Nomogram for power for Vertical Roller Mills when grinding materials

of different HGIs to same product fineness for different thruputs

Scale : A :1 cm = 1 tph thruput

         B : 1 cm = 10 KW power @ mill shaft

DEOLALKAR CONSULTANTS

Nomo no.:	4-2-1 page 2
Title:	Nomogram for power for Vertical Roller Mills when grinding materials of different HGIs to the same product fineness for different thruputs
Useful for:	arriving at power of a vertical mill (ball/roller) when grinding coal/other materials of different hardness expressed in H.G.I.s (Hardgrove Index)
Inputs:	1 thruput in tph 2 H.G.I.s 3 Idle power 40 kw 4 Product fineness 70 % passing / 200 mesh
Output:	mill power in kw
Scale:	A : 1 cm = 1 tph B : 1 cm = 10 kw
How to use:	**Example:** draw a line from thruput (15) on line A thru given H.G.I. (50) on line 0-40 and extend to line B and read power of mill (125 kw) on line B

Note:	In a vertical mill, sp. power is inversely proportional to the HGI. When grinding to a fineness of 70% passing 200 mesh,
	sp.power for HGI, 45 = 6.1 kwh/ton
	                       50 = 5.5 kwh/ton
	                       55 = 5 kwh/ton
	                       60 = 4.5 kwh/ton
	50 HGI is assumed basic for all vertical mills

Source:	constructed

A : 1 cm = 1 % product fineness

B : specially constructed

P1 : point of reference

example:

fineness - 65% thru 200 mesh (A)

multiplying factor : 1.05 (B)

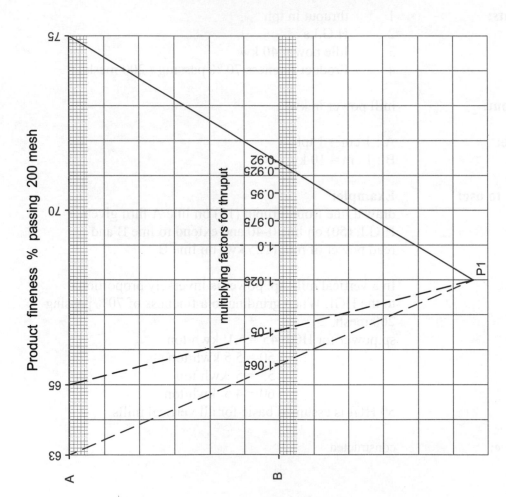

Product fineness % passing 200 mesh

multiplying factors for thruput

Nomogram for multiplying factors for deviations from
base parameters like product fineness & H.G.I.

Scale C : specially constructed

D : 1 cm = 5 HGI

example:

H.G. Index = 60 (D)

multiplying factor : = ~ 1.18 (C)

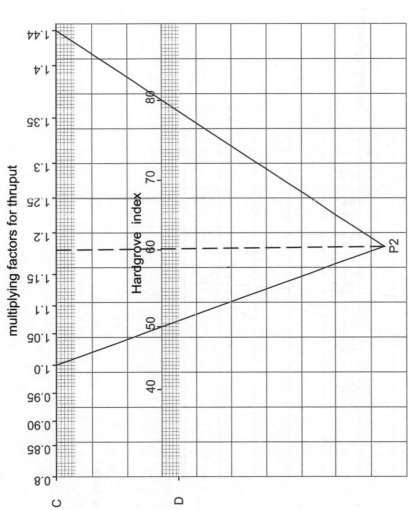

Nomogram for multiplying factors for deviations from base parameters like product fineness & H. G. I.

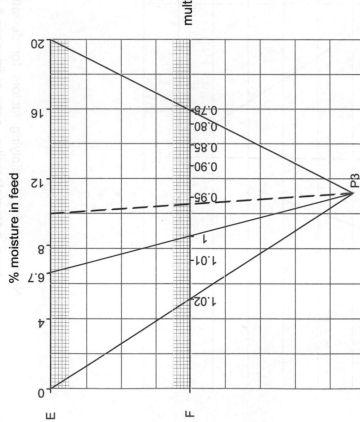

multiplying factor for thruput

% moisture in feed

E

F

P3

scale:

E : 1 cm = 2%
F : specially constructed scale
P3 : point of reference

example:

moisture feed = 10% (line E)

multiplying factor = ~ 0.96 (line F)

Nomogram for multiplying factors for different
moisture content in feed compared to base conditions

DEOLALKAR CONSULTANTS

Nomo no.:	4-2-2 page 4
Title:	Nomogram for multiplying factors (mf) for deviations from base parameters like
	1 product fineness in % passing 200 mesh
	2 HGI
	3 % moisture in feed
Useful for:	arriving at thruput of a vertical mill under actual conditions
Inputs:	1 product fineness in % passing 200 mesh
	2 HGI
	3 Moisture in feed in %
	Note : base parameters are :
	1 product fineness – 70 % passing 200 mesh
	2 HGI - 50
	3 Moisture in feed – 6.7 %
Outputs:	multiplying (correcting) factors (mf) to obtain correct thruput under actual conditions
Scale:	page 1 multiplying factor for thruput on account of product fineness (mf1)
	A: 1 cm = 1 % product fineness
	B : specially constructed scale for multiplying factor
	P1 : point of reference
	page 2 multiplying factor for thruput on account of HGI
	C : 1 cm = specially constructed scale for multiplying factor (mf2)
	D : 1 cm = 5 HGI
	P2 : point of reference
	page 3 multiplying factor on account of % moisture in feed (mf3)
	E : 1 cm = 2 % moisture
	F : specially constructed scale for multiplying factor
	P3 : point of reference

4-2-2   page 5

How to use: In each case, draw a line from the point of reference,
(P1, P2 or P3 as the case may be) to join the respective
variable on lines A, D or E as the case may be and read
pertinent multiplying factor on line B, C, or F.
Total multiplying factor to account for all deviations
would be a product of all multiplying factors

Example:
1   mf1
       product fineness = 65 % passing 200 mesh (line A)
       mf1   = 1.05 (line B)

2   mf2
       HGI   = 60 (line D)
       mf2   = ~ 1.18 (line C)

3   mf3
       % moisture = 10 (line E)
       mf3   = ~ 0.96 (line F)

Total multiplying factor = mf1 × mf2 × mf3
                         = 1.05 × 1.18 × 0.96 = ~1.19

Source: nomogram constructed

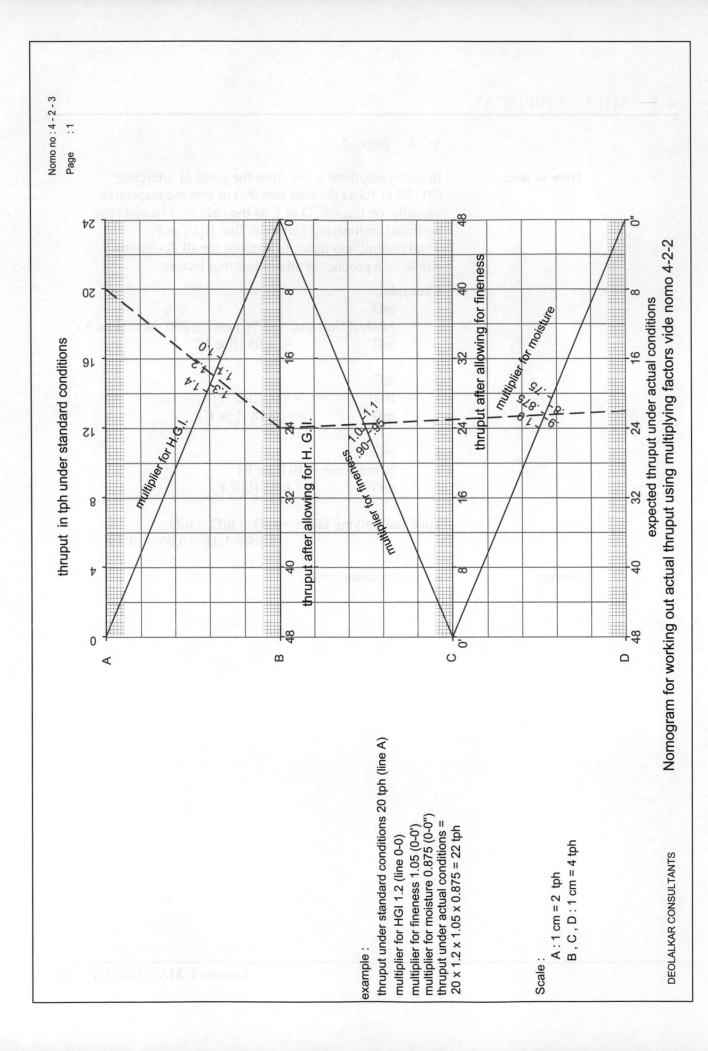

thruput in tph under standard conditions

multiplier for H.G.I.

multiplier for H. G.I.

thruput after allowing for H. G.I.

multiplier for fineness

thruput after allowing for fineness

multiplier for moisture

thruput after allowing for fineness

expected thruput under actual conditions

Nomogram for working out actual thruput using multiplying factors vide nomo 4-2-2

example :
thruput under standard conditions 20 tph (line A)
multiplier for HGI 1.2 (line 0-0)
multiplier for fineness 1.05 (0-0')
multiplier for moisture 0.875 (0-0")
thruput under actual conditions =
20 x 1.2 x 1.05 x 0.875 = 22 tph

Scale :
A : 1 cm = 2 tph
B , C , D : 1 cm = 4 tph

DEOLALKAR CONSULTANTS

Nomo no.:	4-2-3 page 2
Title:	Nomogram to work out actual thruput by using multiplying factors vide nomo 4-2-2
Useful for:	finding out actual thruput or effect on it should parameters change
Inputs :	1 thruput under base conditions 2 actual conditions w.r.t. H.G.I., fineness of product and moisture in feed 3 corresponding multiplying factors obtained from nomo 4-2-2
Output:	thruput under actual conditions
Scale:	A : 1 cm = 2 tph B : 1 cm = 4 tph C : 1 cm = 4 tph D : 1 cm = 4 tph
How to use:	**Example:** draw a line from thruput under base conditions on line A (20) thru multiplying factor for HGI on line 0-0 (1.2) to meet line B (24). From this point draw a line thru multiplying factor for fineness on line 0-0' (1.05) and extend to meet line C (25.2) From this point draw a line thru multiplying factor for moisture on line 0'-0'' (0.875) to meet line D and read thruput under actual conditions (22)
Source:	constructed

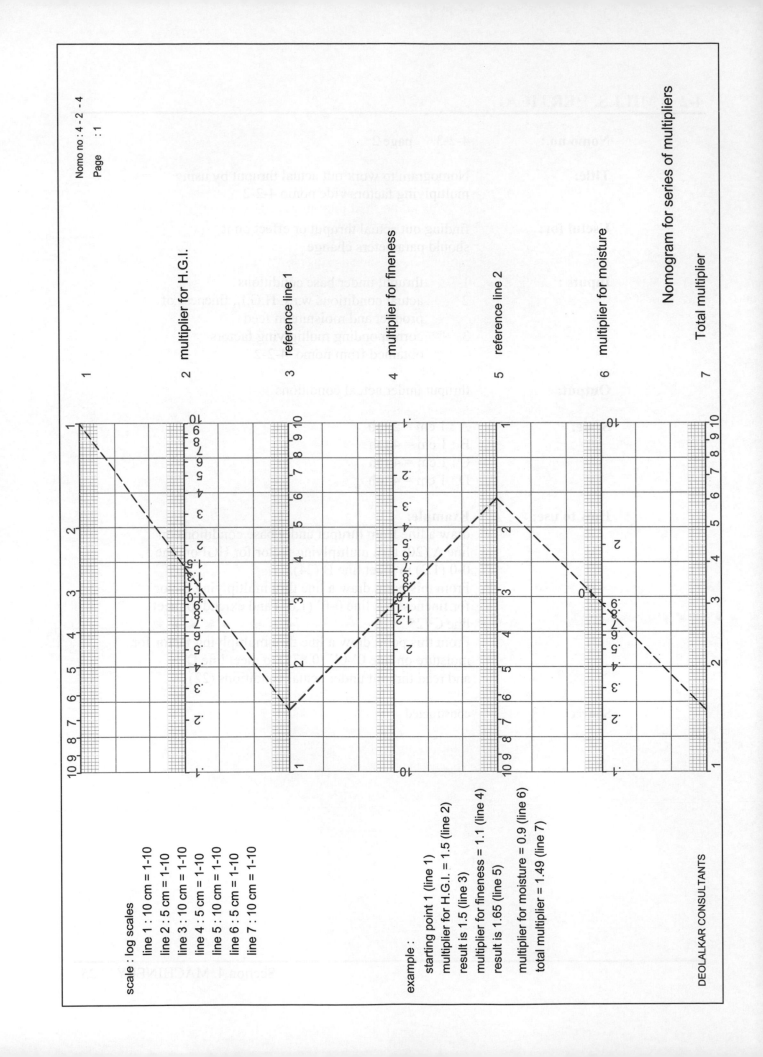

Nomo no : 4 - 2 - 4

Page   : 1

1

2   multiplier for H.G.I.

3   reference line 1

4   multiplier for fineness

5   reference line 2

6   multiplier for moisture

7   Total multiplier

Nomogram for series of multipliers

scale : log scales
    line 1 : 10 cm = 1-10
    line 2 : 5 cm = 1-10
    line 3 : 10 cm = 1-10
    line 4 : 5 cm = 1-10
    line 5 : 10 cm = 1-10
    line 6 : 5 cm = 1-10
    line 7 : 10 cm = 1-10

example :
    starting point 1 (line 1)
    multiplier for H.G.I. = 1.5 (line 2)
    result is 1.5 (line 3)
    multiplier for fineness = 1.1 (line 4)
    result is 1.65 (line 5)
    multiplier for moisture = 0.9 (line 6)
    total multiplier = 1.49 (line 7)

DEOLALKAR CONSULTANTS

Nomo no.:	4-2-4 page 2
Title:	Nomogram for series of multipliers
Useful for:	arriving at the final total multiplier for direct computation of final output
Inputs:	multipliers of various stages
Output:	total multiplier

Note: In nomo 4-2-3, where there were 3 variables, final thruput (result) was obtained in 3 steps.
In this nomogram final multiplier would be obtained by multiplying factors of each stage and final thruput is obtained in one step.

Scale:	there are 7 parallel lines

line 1 : log scale :10 cm = 1-10
line 2 : log scale : 5 cm = 1-10 multiplier for HGI
line 3 : reference line 1, 10 cm = 1-10
line 4 : log scale : 5 cm = 0.1 –1 multiplier for fineness
line 5 : reference line 2, 10 cm = 1-10
line 6 : 5 cm = 1-10 multiplier for moisture
line 7 : 10 cm = 1-10 total multiplier

How to use:

Example:
begin with point 1 on line 1 representing base or standard conditions
Draw line thru factor (1.5) for HGI on line 2 and extend to meet line 3 (reference line 1).
From this point draw line thru multiplying factor (1.1) for fineness on line 4 to meet line 5 (reference line 2).
From this point draw a line thru multiplying factor for moisture (0.9) on line 6 to meet line 7 and read total multiplying factor (1.49).
Thruput under actual conditions would be =
Thruput under base conditions × total multiplying factor
Therefore if base thruput is 20 tph, thruput under actual conditions = 20 × 1.49 = 29.8 tph

Source:	constructed

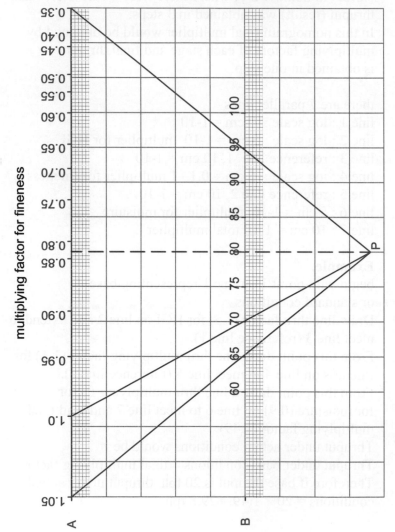

multiplying factor for fineness

% passing 200 mesh
75 microns

scale:

A - specially constructed scale

B - 1 cm = 5%

P - point of reference

example:

% passing 200 mesh - 80% (line B)

multiplying factor = 0.82 (line A)

Nomogram for multiplying factors for fineness for
thruput of vertical E- mill.

Nomo no.:	4-2-5 page 2
Title:	Nomogram for finding out multiplying factor for fineness for thruput of vertical E -mill
Useful for:	arriving at thruput under actual conditions
Inputs:	% passing 200 mesh material coal
Output:	multiplying factor
Scale:	A : specially constructed scale B : 1 cm = 5 % P : point of reference
How to use:	draw a line from P thru required fineness on line B to meet line A and read multiplying factor. **Note:** base condition is 70 % passing 200 mesh when multiplying factor is 1
	Example: Fineness – 80 % passing 200 mesh (line B) multiplying factor = ~ 0.82 (line A)
Source:	base data from Babcock and Wilcox nomogram constructed

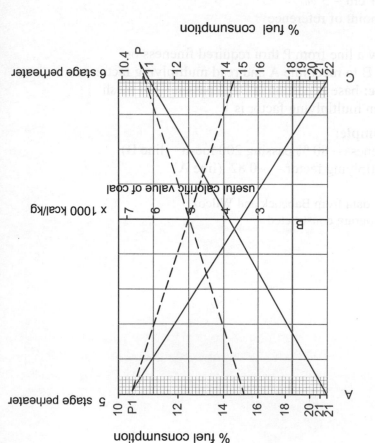

example-

6 Stage preheater - point of reference P1

calorific value of coal : 5000 kcal / kg (line B)

% fuel consumption : 14.4 (line C)

5 Stage preheater - point of reference P

calorific value of coal : 5000 kcal / kg (line B)

% fuel consumption : 15.2 (line A)

Scale : A : Specially constructed scale for % fuel for 5 stage preheater

B : 1 cm =1000 kcal/kg

C : Specially constructed scale for % fuel for 6 stage preheater

P : Point of reference for 5 stage preheater

P1 : Point of reference for 6 stage preheater

Nomogram for sp. fuel consumption in 5 & 6 stage preheaters

for different calorific values of coal

DEOLALKAR CONSULTANTS

Nomo no.:	4-3-1 page 2
Title:	Nomogram for sp. fuel consumption in 5 and 6 stage preheaters for different calorific values of coal
Useful for:	heat balances and for designing coal grinding and firing systems
Inputs:	1 useful calorific values of coal in kcal/kg (3000-7000)
Outputs:	1 sp. fuel consumption for 5 stage preheater in % fuel consumed 2 sp. fuel consumption for 6 stage preheater in % fuel consumed
Scale:	A : specially constructed scale for % fuel for 5 stage preheater B : 1 cm = 1000 kcal/kg P : point of reference C : specially constructed scale for % fuel for 6 stage preheater P1 : corresponding point of reference
How to use:	**Example:** 1 6 stage preheater draw a line from point of reference, P1 thru calorific value of coal on line B (5000) and extend to line C and read % fuel consumption (14.4 %) 2 5 stage preheater draw a line from reference point P thru calorific value of coal on line B (5000) and extend it to line A and read % fuel consumption (15.2 %)
Source:	constructed

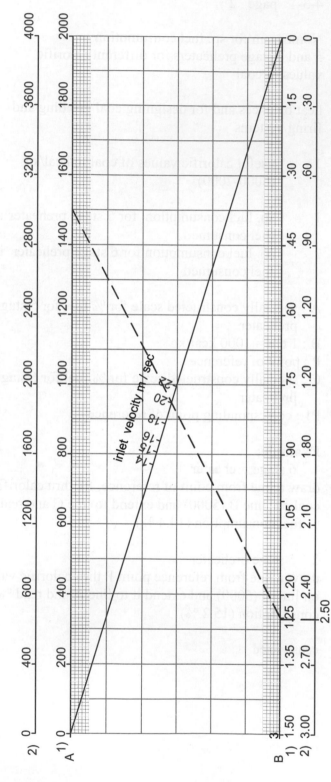

Gas flow m³ / min

inlet velocity m / sec

inlet area of cyclone in m²

example :

Gas flow = 1500 ³ m / min (line A)

Vel.@ inlet = 20 m/sec   (line 0-0)

area of cyclone inlet = 1.25 m² (line B)

Nomogram for finding inlet areas of cyclones for dust collectors and preheaters

scale :

A :1 cm =100 / 200 m ³ / min

B :1 cm = 0.075m / 0.15 m²

DEOLALKAR CONSULTANTS

Nomo no.:	4-3-2 page 2
Title:	Nomogram for finding inlet areas of cyclones for dust collectors and preheaters
Useful for:	Sizing cyclones for dust collection systems and for preheaters
Inputs:	1 gas flow in m³/min. 2 velocity of gases at entrance in m³/sec.

Output: in Part 1   area of inlet of cyclone

Note : Same nomogram can be used for double the gas flow :
1   inlet area is doubled size of cyclone changes accordingly
2   no. of cyclones doubled, same size

Scale: A : 1cm = 100 / 200 m³/min
B : 1 cm = 0.075 / 0.15 m²

How to use: draw a line from gas flow in m³/min on line A to pass thru, point showing velocity on line 0-0 and extend to meet line B. Read inlet area of cyclone on it.

Example:
Gas flow = 1500 m³/min (line A 1)
Velocity at inlet = 20 m/sec (line 0-0)
Inlet area= ~ 1.25 m² (line –B 1)
Gas flow = 3000 m³/min (line A 2)
Velocity = 20 m/sec (line 0-0)
Inlet area = 2.5 m² (line –B 2)
In case, 2 cyclones are used in parallel, size of cyclone would remain same

Source: constructed

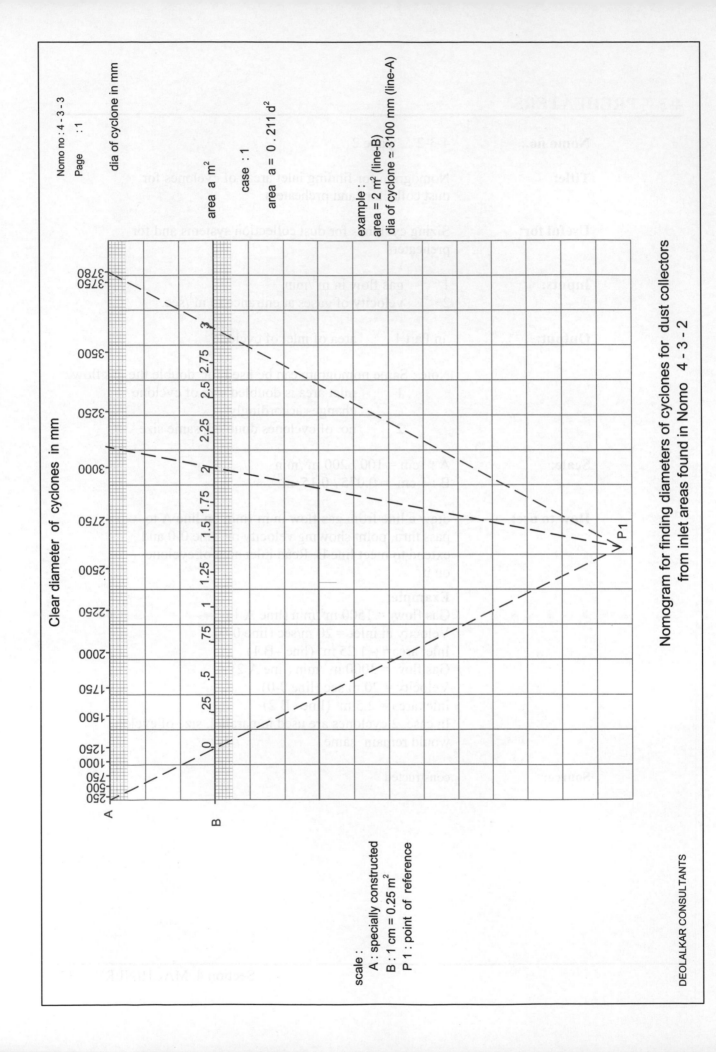

Nomo no : 4 - 3 - 3
Page   : 1

dia of cyclone in mm

area a m²

case : 1

area a = 0. 211 d²

example :
area = 2 m² (line-B)
dia of cyclone ≈ 3100 mm (line-A)

Clear diameter of cyclones in mm

scale :
A : specially constructed
B : 1 cm = 0.25 m²
P 1 : point of reference

Nomogram for finding diameters of cyclones for dust collectors
from inlet areas found in Nomo 4 - 3 - 2

DEOLALKAR CONSULTANTS

Case 2

$a = 0.125 \, d^2$

applicable to top cyclones

of preheaters

Clear diameter of cyclone in mm

Inlet area   a = m²

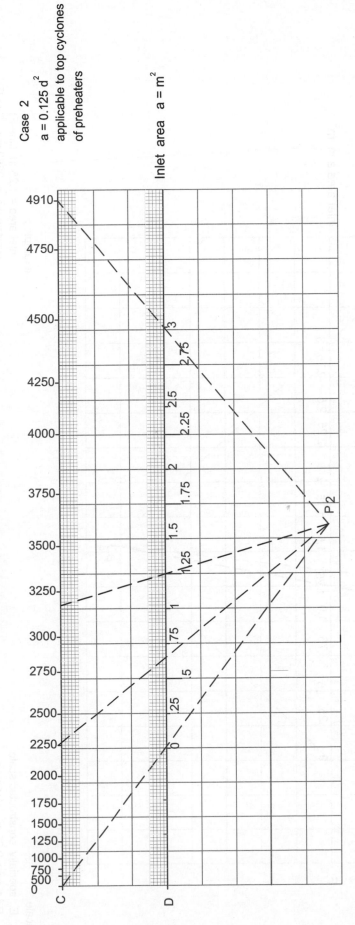

example :

   Inlet area = 1.25 m² (line-D)

   cyclone dia = 3200 mm (line-C)

   If 2 cyclones in stage,

   inlet area = 0.625 m²

   cyclone dia = ~ 2250 mm

scale :

  C = specially constructed scale

  D = 1 cm = 0.25 m²

  P 2 = point of reference

Nomogram for finding diameters of top cyclones of preheater

from inlet areas found in Nomo 4 - 3 - 2

DEOLALKAR CONSULTANTS

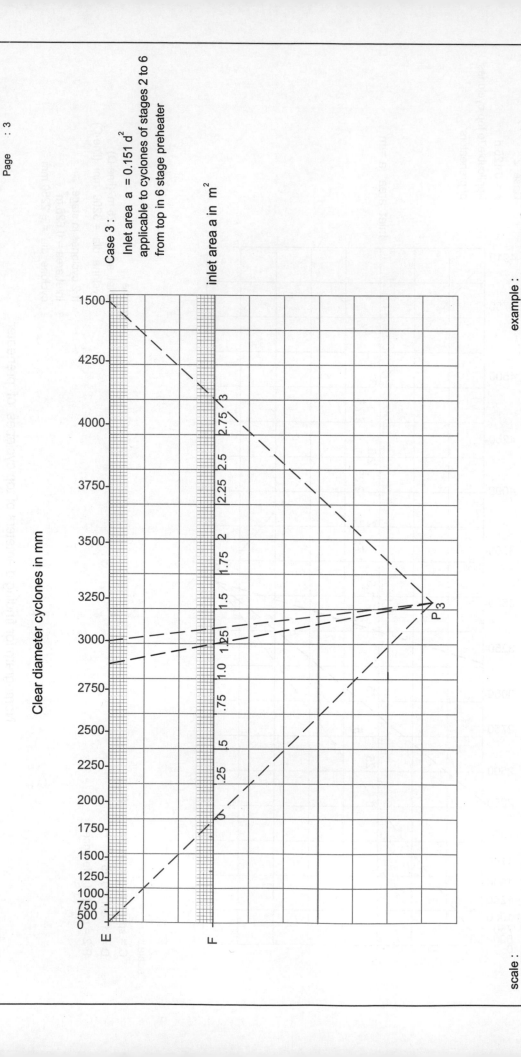

Case 3 :

Inlet area a = 0.151 d^2
applicable to cyclones of stages 2 to 6
from top in 6 stage preheater

inlet area a in m^2

Clear diameter cyclones in mm

1500

4250

4000

3750

3500

3250

3000

2750

2500

2250

2000

1750

1500
1250
1000
750
500
0

3

2.75

2.5

2.25

2

1.75

1.5

1.25

1.0

.75

.5

.25

0

P3

E

F

example :

Inlet area = 1.25 m^2 (line-F)
cyclone dia ≃ 2850 mm (line-E)

Nomogram for finding diameters of cyclones of other stages

of preheater from inlet area found in Nomo 4-3-2

scale :

E : specially constructed scale

F : 1 cm = 0.25 m^2 inlet area

P3 : Point of reference

DEOLALKAR CONSULTANTS

Nomo no.:	4-3-3 page 4
Title:	Nomogram for finding diameters of cyclones of dust collectors and preheaters from inlet areas found in Nomo 4-3-2
Useful for:	sizing cyclones for primary dust collector systems and for preheaters
Note:	In nomo 4-3-2, inlet area of cyclone was worked out from gas flow and velocity at inlet. In this nomo, diameter of cyclone would be worked out from different designs of cyclones
Input:	area of inlet of cyclone from nomo 4-3-2
Output:	size of cyclone in terms of diameter other dimensions can be arrived at from relative proportions w.r.t. diameter in different designs
Note:	In case of preheater cyclones where gas temperatures are more than 300 $^\circ$ C, inlet areas would be mean clear area inside of bricks and diameters would be clear inside of bricks

Scale:

page 1   case 1   cyclones for dust collectors
$$a = 0.211 \times d^2$$
(a = inlet area, d = diameter of cyclone)
A : specially constructed scale
B : 1 cm = 0.25 m^2
Page 2   case 2   top cyclones of preheater
$$a = 0.125 \times d^2$$
C : specially constructed scale
D : 1 cm = 0.25 m^2
Page 3   case 3 cyclones in other stages of preheater
$$a = 0.151 \times d^2$$
E : specially constructed scale
F :   1 cm = 0.25 m^2
P1, P2, & P3 are points of references respectively for 3 cases

Nomo no.:	4-3-3 page 5
How to use:	select design of cyclone (case 1, 2 or 3). case 1 is for general primary dust collector system of cyclones case 2 is for top cyclones of a preheater system and case 3 is for other stages of the same system In each case draw line from the respective reference point P thru corresponding lines B,D or F to meet corresponding lines A, C or E and read clear diameter of cyclone in mms.

Example:
Case 1
Inlet area = 2 m^2 (line B)
Cyclone dia. = ~3100 mm (line A)
Case 2
Inlet area = 1.25 m^2 (line D)
Cyclone dia = ~ 3200 mm (line C)
If top stage has two cyclones, then inlet area
Each cyclone = 0.625 m^2 and its dia would be
Found out in same fashion = ~ 2250 mms
Case 3
Inlet area = 1.25 m^2 (line F)
Cyclone dia = ~ 2850 mm (linc E)

Source:	constructed

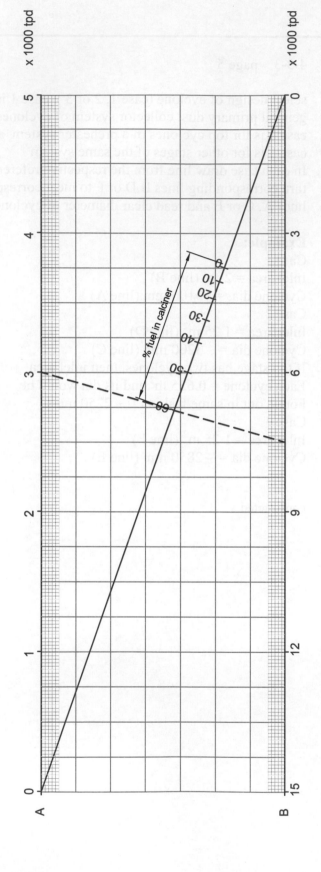

x 1000 tpd

x 1000 tpd

Capacity of preheater kiln tpd

% fuel in calciner

Resultant capacity of calciner kiln in tpd for
different % fuel fired in calciner

example : Capacity preheater kiln 3000 tpd (line A)
          % fuel in calciner 60 (line 0-0)
          capacity calciner 7500 tpd (line B)

Nomogram for working out increased capacity of a
preheater kiln for different percentages of fuel fired in calciner

Scale : A :1 cm = 250 tpd - capacity preheater kiln
        B : 1 cm = 750 tpd - capacity of calciner kiln

DEOLALKAR CONSULTANTS

Nomo no.:	4-4-1 page 2
Title:	Nomogram for working out increased capacity of a prheater kiln for different percentages of fuel fired in calciner
Useful for:	working out resultant capacity of a calciner kiln and for system design
Inputs:	1 capacity of preheater kiln in tpd (rated or design) 2 % of fuel fired in calciner out of total fuel
Output:	resultant capacity in tpd (rated or design)
Scale:	A : 1 cm = 250 tpd B : 1 cm = 750 tpd
How to use:	draw a line from capacity of preheater kiln on line A through % fuel online 0-0 to met line B and read resultant capacity of calciner kiln

Example:
capacity of preheater kiln = 3000 tpd (line A)
% fuel in calciner = 60 (line 0-0)
capacity of calciner kiln = 7500 tpd (line B)

Source:	constructed

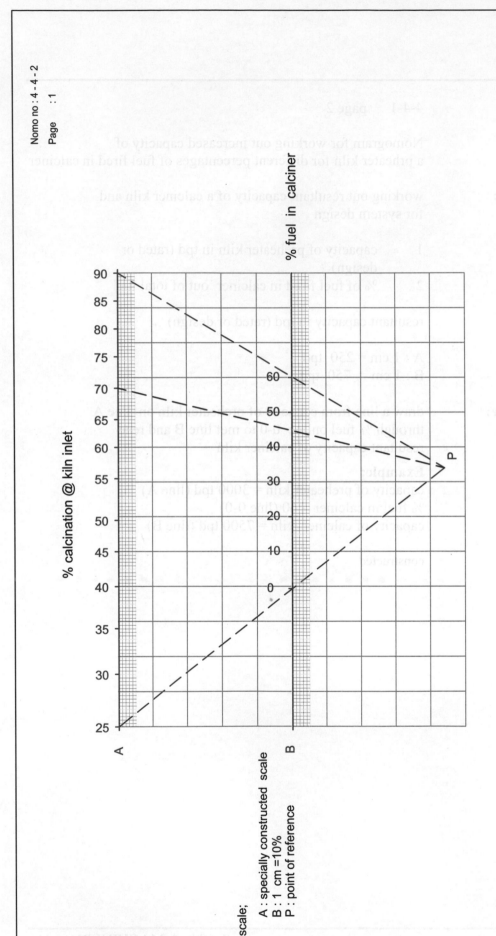

% calcination @ kiln inlet

% fuel in calciner

scale;

A : specially constructed scale
B : 1 cm = 10%
P : point of reference

Nomogram for percentage fuel fired in calciner and degree of calcination at kiln inlet

DEOLALKAR CONSULTANTS

Nomo no.:	4-4-2 page 2
Title:	Nomogram for percent fuel fired in calciner and degree of calcination at kiln inlet
Useful for:	determining capacity of kiln
Inputs:	% fuel in calciner
Output:	degree of calcinations in % at kiln inlet
Scale:	A : specially constructed scale B : 1 cm = 10 % P : point of reference
How to use:	**Example:** Draw a line from point of reference P thru % fuel in calciner on line B(45) and extend it to meet line A and read degree of calcination (70)%
Source:	Mitsubishi Manual nomogram constructed

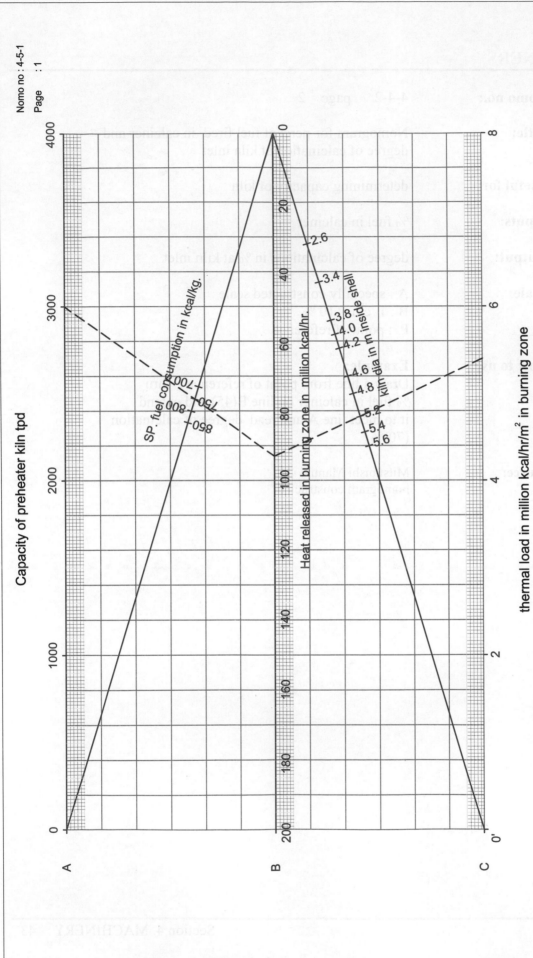

Capacity of preheater kiln tpd

SP. fuel consumption in kcal/kg.

850 800 750 700

Heat released in burning zone in million kcal/hr.

kiln dia in m inside shell

2.6
3.4
3.8
4.0
4.2
4.6
4.8
5.2
5.4
5.6

thermal load in million kcal/hr/m² in burning zone

example : kiln capacity tpd, 3000 (line A), sp. fuel consumption 750 kcal/kg. (line 0-0)
heat released (line B), 93.75 million kcal/hr.
dia of kiln 5.2, (line 0-0'), thermal load 5.4 million kcal/hr/m²

Nomogram for working out thermal load in million kcal/hr/m² in the burning zone
in preheater kilns

Scale : A :1cm = 200 tpd
         B : 1cm = 10 million kcal/hr
         C : 1cm = 0.4 million kcal/hr/m²

DEOLALKAR CONSULTANTS

Nomo no.:	4-5-1 page 2
Title:	Nomogram for working out thermal load in million kcal/hr/m^2 in the burning zone in preheater kilns
Useful for:	selecting refractories in the burning zone
Inputs:	1 capacity of preheater kiln in tpd 2 sp. fuel consumption in kcal/kg clinker 3 nominal dia of kiln in metres
Output:	1 heat released in million kcal/hr 2 thermal load in million kcal/hr/ m^2
Scale:	A : 1 cm = 200 tpd B : 1 cm = 10 million kcal/hr C : 1 cm = 0.4 million kcal/hr/m^2
How to use:	From point showing kiln capacity on line A, draw line thru sp. fuel consumption on line 0-0 and extend to meet line B. read heat released in million kcal/hr. From this point draw line thru dia. of kiln on line 0-0' and read thermal load in million kcal/hr/m^2

Example:
Kiln capacity :   3000 tpd (line A)
Sp. fuel consumption : 750 kcal/kg (line 0-0)
Heat released in burning zone :
~ 93.75 million kcal/hr (line B)
Dia. of kiln :   5.2 m inside shell (line 0-0')
Thermal load in burning zone :
$\simeq$ 5.4 million kcal/hr/m^2 (line C)

Note:	when the kiln is used as a precalciner kiln, heat released in burning zone remains the same whatever % fuel in calciner and consequent output of kiln
Source:	constructed

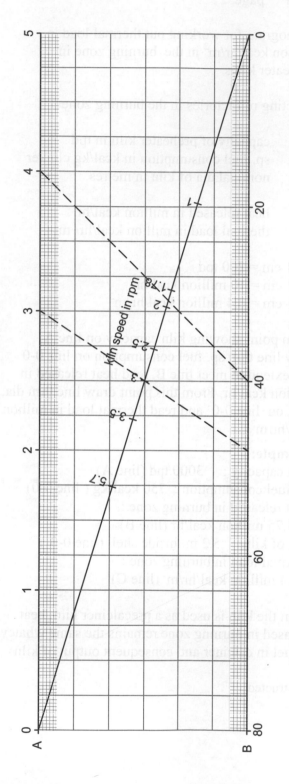

Kiln diameter in metres

Kiln speed in rpm

Kiln linear speed in cm / sec

example : 1 kiln dia. 3 m, (line A), rpm 3 (line 0-0), linear speed ~ 48 cm / sec (line B)

2 kiln dia. 4 m, (line A), linear speed, 40 cms / sec (ine B), rotating speed ~1.9 rpm (0-0)

**Nomogram for working out linear speed of rotary kilns, coolers etc.,
of different diameters running at different speeds in r.p.m. and vice versa**

Scale : A 1 cm = 0.25 m dia kiln

B 1 cm = 4 cm / sec

DEOLALKAR CONSULTANTS

Nomo no.:	4-5-2 page 2
Title:	Nomogram for working out linear speeds of rotary kilns, coolers etc., of different diameters running at different speeds in r.p.m. and vice versa
Useful for:	converting linear speeds into r.p.m. and vice versa for sizing gearboxes
Inputs:	1 nominal diameters of kiln, cooler etc. 2 rotational speeds in r.p.m.
Output:	linear kiln speed in cm/sec.
Scale:	A : 1 cm = 0.25 m dia of kiln B : 1 cm = 4 cm/sec linear speed
How to use:	Draw a line from given kiln diameter on line A thru kiln speed in r.p.m. and extend it to line B, and read linear kiln speed in cm / sec. Same nomogram can be used to convert linear speeds into rotational speeds in r.p.m.

Example:
1   Kiln dia. : 3 m (line A)
     Kiln speed : 3 r.p.m. (line 0-0)
     Linear speed = ~ 48 cm / sec (line B)
2   Kiln dia : 3 m (line A)
     Linear speed : 40 cm/sec (line B)
     Rotating speed = ~ 1.9 r.p.m.

Source:	constructed

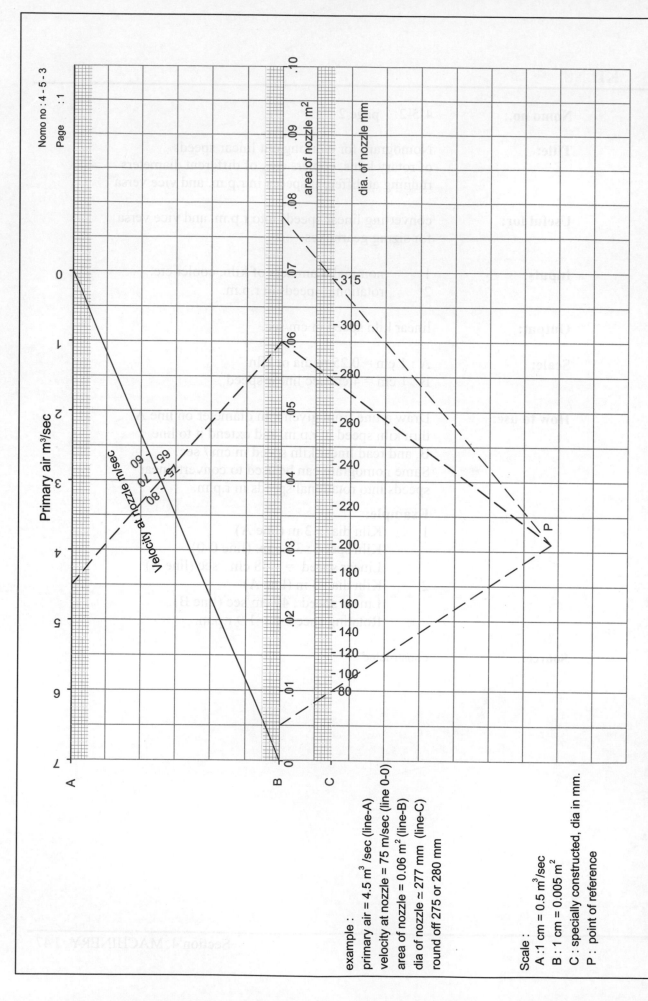

Nomo no : 4 - 5 - 3
Page : 1

Primary air m³/sec

area of nozzle m²

dia. of nozzle mm

Velocity at nozzle m/sec

A   B   C   P

example :
primary air = 4.5 m³ /sec (line-A)
velocity at nozzle = 75 m/sec (line 0-0)
area of nozzle = 0.06 m² (line-B)
dia of nozzle ≃ 277 mm (line-C)
round off 275 or 280 mm

Scale :
A :1 cm = 0.5 m³/sec
B : 1 cm = 0.005 m²
C : specially constructed, dia in mm.
P : point of reference

DEOLALKAR CONSULTANTS   Nomogram for working out dia of nozzle of coal firing pipe in kiln, knowing quantity of primary air & velocity at nozzle

Nomo no.:	4-5-3 page 2
Title:	Nomogram for working out diameter of nozzle of coal firing pipe in kiln knowing quantity of primary air and velocity at nozzle
Useful for:	arriving at nozzle size and hence dimensions of coal firing pipe
Inputs:	1 quantity of primary air in kin in m^3/sec
	2 velocity at nozzle which varies between 60 to 80 m/sec depending on diameter of kiln. Quantity of primary air is dependent on design of burner used and varies between 12 to 20 % of total air for combustion
Output:	diameter of nozzle in mm
Scale:	A : 1 cm = 0.5 m^3/sec primary air
	B : 1 cm = 0.05 m^2 area of nozzle
	C : specially constructed scale for dia. of nozzle in mm
	P : reference point
How to use:	From point of quantity of primary air on line A draw a line through velocity of nozzle on 0-0 to meet line B. Join this point with reference point P and read dia. of nozzle on line C

Example:
Primary air = 4.5 m^3/sec   (line A)
Velocity at nozzle = 75 m/sec (line 0-0)
Area of nozzle   = 0.06 m^2 (line B)
Dia. of nozzle = ~277 mm (line C)
Round off to 275 / 280 mm

Source:	constructed

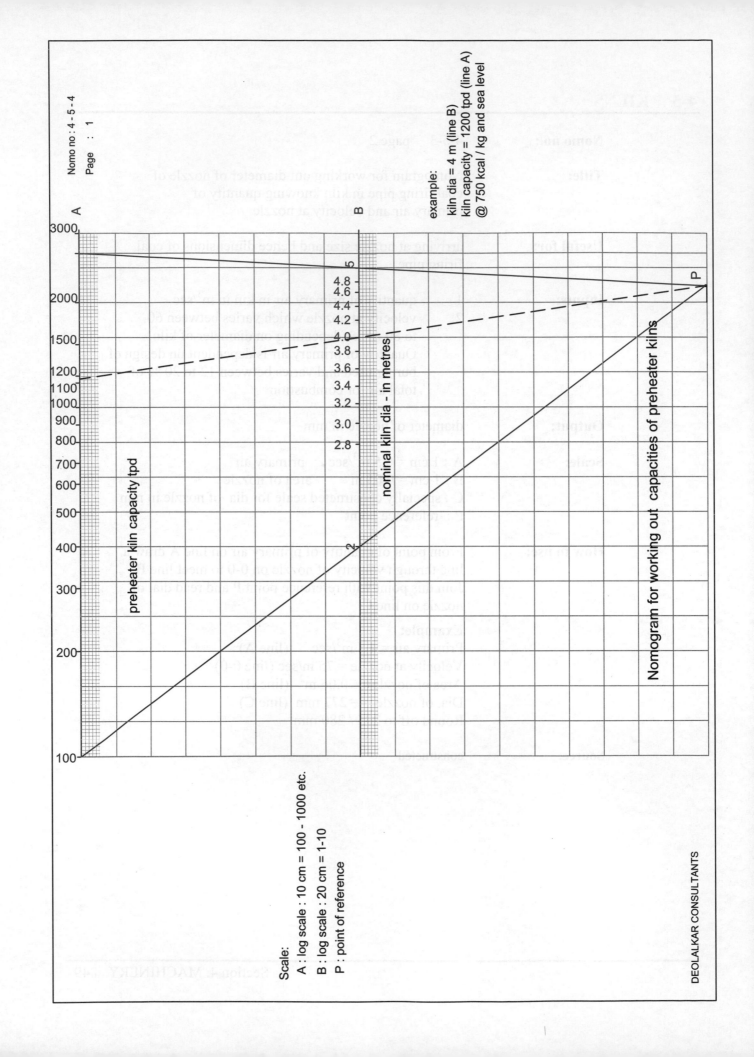

example:
kiln dia = 4 m (line B)
kiln capacity = 1200 tpd (line A)
@ 750 kcal / kg and sea level

A

B

P

3000

2000

1500

1200
1100
1000
900
800
700
600
500

400

300

200

100

5

4.8
4.6
4.4
4.2
4.0
3.8
3.6
3.4
3.2
3.0
2.8

2

preheater kiln capacity tpd

nominal kiln dia - in metres

Nomogram for working out capacities of preheater kilns

Scale:

A : log scale : 10 cm = 100 - 1000 etc.
B : log scale : 20 cm = 1-10
P : point of reference

DEOLALKAR CONSULTANTS

Nomo no.:	4-5-4 page 2
Title:	Nomogram for working out capacities of preheater kilns
Useful for:	quick estimation of capacities of preheater kilns
Inputs:	1 nominal dia. of kiln in metres
	2 basic assumptions
	altitude – sea level
	sp. fuel consumption –750 kcal/kg
	L/D ratio –15: 1 (length /diameter ratio)
Output:	capacity of kiln in tpd (tons per day clinker)
Note:	based on Mitsubishi formula : $C = 8.17 \times D^{3.62}$
Scale:	A : log scale – 10 cm =100-1000 etc.
	B : log scale -- 20 cm = 1-10
	P : point of reference

How to use:

From reference point P, draw a line thru dia. of kiln on line B and extend to met line A and read capacity of preheater kiln in tpd

For other common L/D ratios, use multiplying factors

	12	14	15	16
multiplier	0.93	0.98	1.0	1.022

For other sp. fuel consumptions use multiplying factors

sp. fuel consumption kcal/kg	700	750	800	850
multiplier	1.07	1	0.94	0.88

Example:
kiln dia = 4 m (line B)
L/D = 15
Sp. fuel consumption = 800 kcal/kg
Kiln capacity = ~ 1200 tpd for sp. fuel consumption
            of 750 kcal/kg   (line A)
multiplying factor for 800 kcal = 0.94
therefore capacity = ~ 1130 tpd
for altitude, capacity would reduce inversely to density of air.

Source: base formula from Mitsubishi manual
nomogram constructed

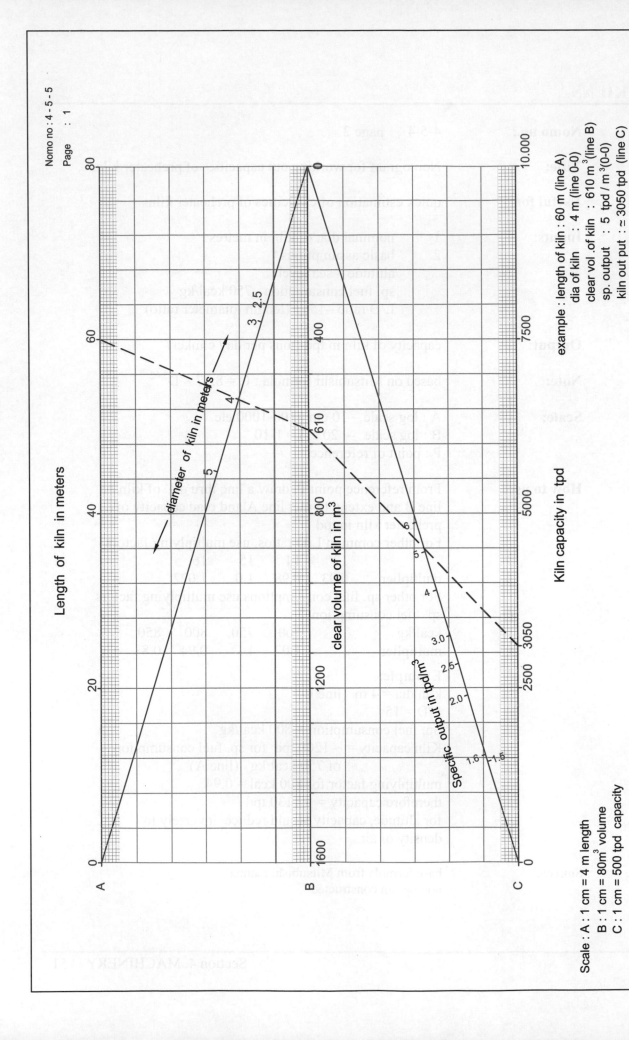

Nomo no : 4 - 5 - 5
Page : 1

example : length of kiln : 60 m (line A)
dia of kiln : 4 m (line B)
clear vol. of kiln : 610 m³ (line 0-0)
sp. output : 5 tpd / m³ (0-0)
kiln out put : ≈ 3050 tpd (line C)

Length of kiln in meters

Kiln capacity in tpd

clear volume of kiln in m³

Specific output in tpd/m³

diameter of kiln in meters

Scale : A : 1 cm = 4 m length
B : 1 cm = 80m³ volume
C : 1 cm = 500 tpd capacity

Nomogram for working out capacity of a rotary kiln from its size & specific output in tpd / m³

DEOLALKAR CONSULTANTS

Nomo no.:	4-5-5 Page 2
Title:	Nomogram for working out capacity of a rotary kiln from its size and specific output in tpd / m^3
Useful for:	Quick estimation of preheater and calciner kilns
Inputs:	1 size of kiln – dia × length
	2 brick thickness mm
	3 sp. output in tpd/m^3 for preheater and calciner kilns
Output:	capacity of kiln in tpd
Scale:	A : 1 cm = 4 m length
	B : 1 cm = 80 m^3 volume
	C : 1 cm = 500 tpd capacity
How to use:	capacity of preheater kilns was obtained in nomo 4-5-4. From it specific output in tpd/m^3 can be derived. It ranges between 1.65 and 2.3 tpd/m^3 at 750 kcal/kg and sea level
	sp. outputs for calciner kilns range between 4 and 6 tpd/m^3 with 60 % fuel in calciner

Example:
kiln 4 m dia × 60 m long
Draw a line from 60 on line A thru 4 on line 0-0
and extend to line B and read clear volume 610 m^3.
From this point draw a line thru point 5 on line 0-0'
and extend to line C and read capacity of calciner
kiln as 3050 tpd

Source:	constructed

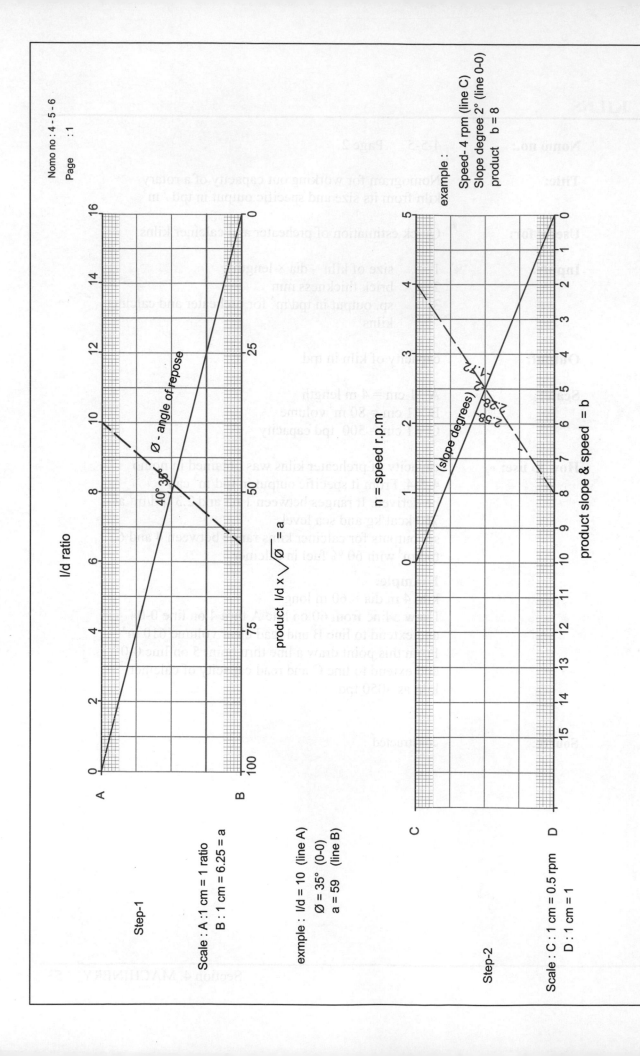

Nomogram for calculating retention time in Rotary Kilns and Coolers

DEOLALKAR CONSULTANTS

Nomo no.:	4-5-6 page 2
Title:	Nomogram for calculating retention time in Rotary Kilns and Coolers
Useful for:	estimating degree of filling and subsequent calculation for power also to see if it is adequate for the process under consideration

Inputs:

1   L/D (length /diameter) ratio   L/D
2   angle of repose in $^\circ$ (normally 35-40 $^\circ$ θ
3   slope in $^\circ$ – ranges from 1.72 to 2.58 $^\circ$ p
4   speed in r.p.m. ranges from 1 to 5 r.p.m.   n
5   multiplier –1 for kilns ; 2- for coolers, dryers   k
6   constant 1.77

formula : retention time t = 1.77 $((L/D) \times (\theta)^{1/2}/(p \times n)) \times k$

Output: retention time in minutes

Nomogram is in 4 steps
Step 1   find product L/D $\times \theta^{1/2}$ = a
Step 2   find product p $\times$ n   = b
Step 3   find a/b
Step 4   use multipliers 1.77 $\times$ 1 or 1.77 $\times$ 2 as
              the case may be.

Scale: steps 1 and 2 have arithmetic scales
              3 and 4 have logarithmic scales

step 1   A : 1 cm = 1 (ratio)
              B : 1 cm = 6.25 = a
step 2   C : 1 cm = 0.5 r.p.m
              D : 1 cm = 1 = b
Step 3   E : log scale 20 cm   =1-100 = a
              F : specially constructed for b
              G : log scale : 10 cm = 1-100 = a/b
              H : log scale 5 cm = 1-10, 10-100 etc.
                              for retention time t
              J : log scale : 10 cm =1-10 = multipliers

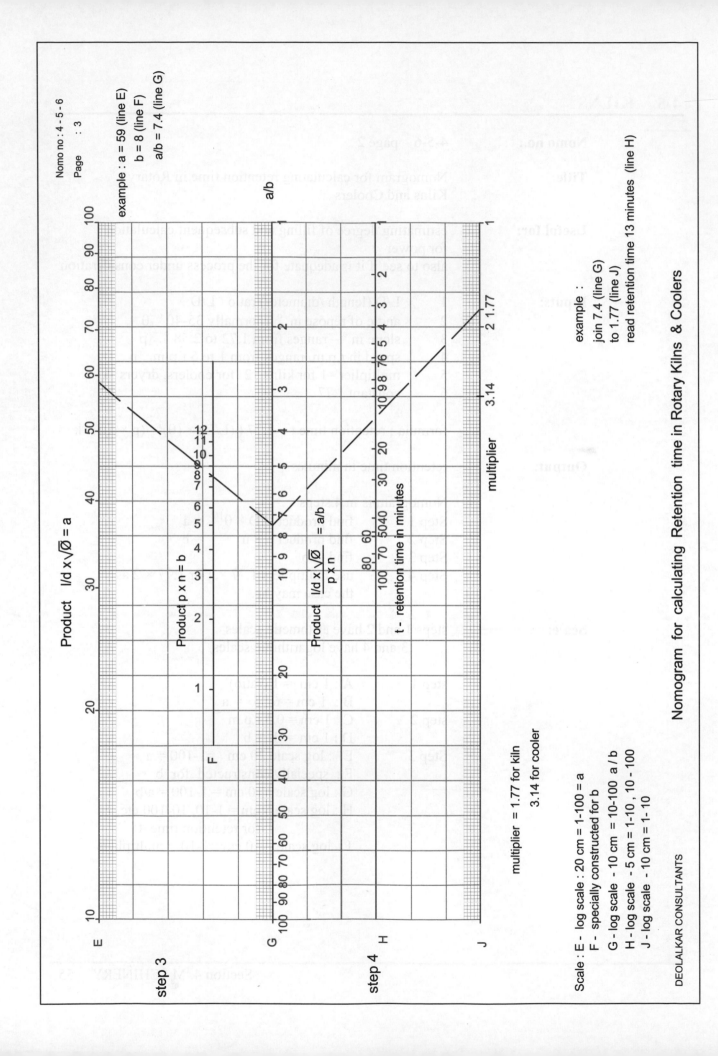

Nomo no : 4 - 5 - 6
Page : 3

example : a = 59 (line E)
b = 8 (line F)
a/b = 7.4 (line G)

example :
join 7.4 (line G)
to 1.77 (line J)
read retention time 13 minutes (line H)

Product l/d x √∅ = a

Product p x n = b

Product (l/d x √∅)/(p x n) = a/b

t - retention time in minutes

multiplier = 1.77 for kiln
3.14 for cooler

Scale : E - log scale : 20 cm = 1-100 = a
F - specially constructed for b
G - log scale - 10 cm = 10-100 a / b
H - log scale - 5 cm = 1-10 , 10 - 100
J - log scale - 10 cm = 1- 10

Nomogram for calculating Retention time in Rotary Kilns & Coolers

DEOLALKAR CONSULTANTS

4-5-6   page   4

How to use:

Example:

Step 1 :

L/D =10; θ, 35 °

draw line from 10 on line A thru 35 °

on line 0-0 to meet line B and read a = 59

Step 2 :

Slope = 3.5% = 2 ° ; speed = 4 r.p.m.

from 4 (n) on line C, draw a line thru

slope p 2 on line 0-0 to meet line D

read b = 8 on it

Step 3 :

From 59 (a) on line E draw a line thru 8

(b) on line F to meet line G in 7.4 = a/b

Step 4 :

Kiln : multiplier = 1.77 × 1

from this point, 7.4 on line G, draw a line

to 1.77 on line J ;

it will cut line H. Read retention time on

it = ~ 13 minutes

For cooler, retention time would be

double = ~ 26 minutes

Source:

Basic formulae from Duda

nomogram constructed

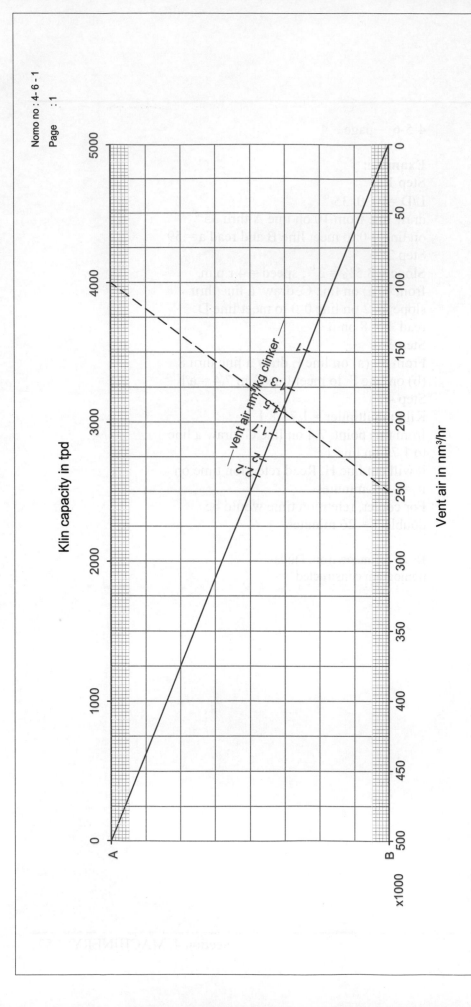

Klin capacity in tpd

Vent air in nm³/hr

vent air nm³/kg clinker

example : kiln cap. 4000 tpd (line A), vent air 1.5 nm³ /kg (line 0-0),
read vent air in nm³ /hr. = 250,000 (line B)

Nomogram for working out quantity of vent air from grate cooler in nm³ /hr.

Scale A: 1 cm = 250 tpd
B: 1 cm = 25000 nm³ / hr

DEOLALKAR CONSULTANTS

Nomo no.:	4-6-1 page 2
Title:	Nomogram for working out quantity of vent air from grate cooler in nm^3/hr
Useful for:	finding out quantity of vent air for sizing dust collector system to handle it
Inputs:	1 kiln capacity in tpd 2 vent air in nm^3/kg clinker vent air = cooling air − (air used as primary + secondary + tertiary air + leakage)
Output:	vent air in nm^3/hr
Scale:	A : 1 cm = 250 tpd B : 1 cm = 25000 nm^3/hr vent air
How to use:	From the point showing capacity of kiln on line A draw a line thru point showing vent air in nm^3/kg clinker on line 0-0 and extend it to meet line B and read vent air in nm^3/hr

Example:

kiln capacity :	4000 tpd (line A)
vent air :	1.5 nm^3/kg (line 0-0)
vent air :	~ 250000 nm^3/hr (line B)

Source:	constructed

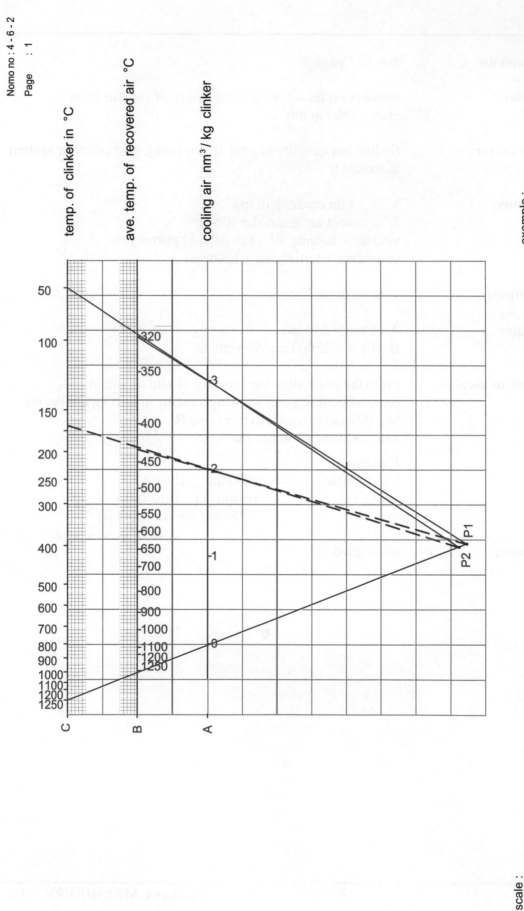

Nomo no : 4 - 6 - 2

Page : 1

temp. of clinker in °C

ave. temp. of recovered air °C

cooling air nm³ / kg clinker

example :

cooling air 2 nm³/kg (line A)
reference point P2
temp of clinker- 170°C (line C)
reference point P1
ave. temp. of recovered air 430°C (B)

**Nomogram for finding temp. of clinker and average temperature
of recovered air in conventional coolers**

scale :

A : 2.5 cm = 1 nm³ /kg cooling air
B : specially constructed scale for temp. recovered air
C : specially constructed scale for temp. of clinker
P1 : point of reference for temp. of recovered air
P2 : point of reference for temp. of clinker

DEOLALKAR CONSULTANTS

Nomo no.:	4-6-2 page 2
Title:	Nomogram for finding temperature of clinker and average temperature of recovered air in conventional coolers .
Useful for:	Heat balance of clinker coolers
Input:	1 cooling air admitted in nm^3/kg clinker 2 temperature of clinker entering cooler 1250 $^\circ$ C
Outputs:	1 average temperature of recovered air in $^\circ$ C 2 temperature of clinker after the cooling air in $^\circ$ C
Scale:	A : 2.5 cm = 1 nm^3/kg of cooling air B : specially constructed scale for ave. temperature of recovered air in $^\circ$ C C : specially constructed scale for temperature of clinker in $^\circ$ C P1 : point of reference for temp. of recovered air P2 : point of reference for temperature of clinker
How to use:	**Example:** 1 temperature of recovered air From point P1 draw a line thru given quantity of cooling air on line A and extend to meet line B and read temperature of recovered air cooling air 2 nm^3/kg (line A) Temp. of recovered air = ≃ 430 $^\circ$ C 2 temperature of clinker From point P2, draw a line thru given quantity of cooling air on line A and extend to meet line C and read temp. of clinker cooling air = 2 nm^3/kg (line A) Temperature of clinker = ≃ 170 $^\circ$ C
Source:	Mitsubishi manual nomogram constructed

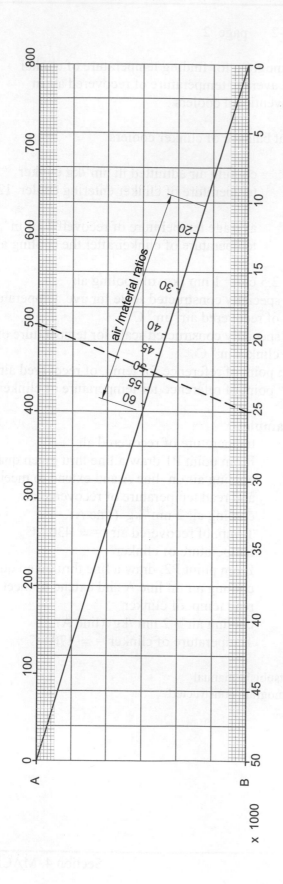

Air lift capacity in tph

conveying air m³ /hr

air /material ratios

Scale : A : 2.5 cm = 100 tph

B : 1 cm = 2500 m³ / hr

Example : capacity : 500 tph (line A)

air / material ratio : 50 line 0-0

conveying air : 25000 m³ /hr (line B)

Nomogram for finding out conveying air in air lifts of

different capacities & air to material ratios

DEOLALKAR CONSULTANTS

Nomo no.:	4-7-1 page 2
Title:	Nomogram for finding out conveying air in m³/hr for air lifts of different capacities and air to material ratios
Useful for:	selecting blowers / compressors in air lift systems and for sizing pipe lines
Inputs:	1 capacity of air lift in tph 2 air to material ratios
Output:	Conveying air in m³/hr
Scale:	A : 2.5 cm = 100 tph B : 1 cm = 2500 m³/hr
How to use:	**Example:** draw a line from given capacity on line A (500 tph) thru given air to material ratio (50) on line 0-0 and read required conveying air on line B in m³/hr (25000)
Source:	base data from CPAG Manual nomogram constructed

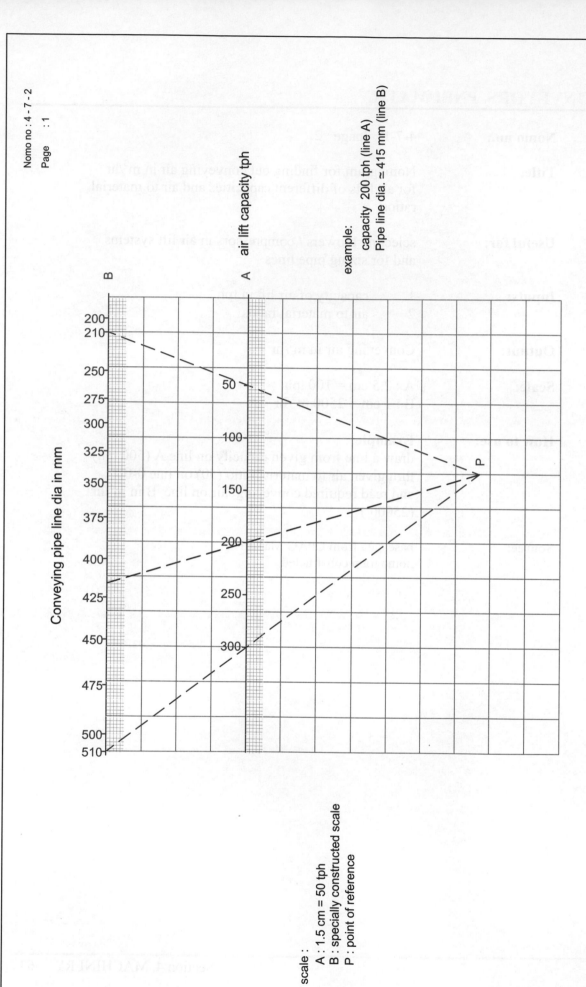

air lift capacity tph

B

A

example:

capacity 200 tph (line A)
Pipe line dia. ≃ 415 mm (line B)

Conveying pipe line dia in mm

200
210

250

275

300

325

350

375

400

425

450

475

500
510

50

100

150

200

250

300

P

scale :
A : 1.5 cm = 50 tph
B : specially constructed scale
P : point of reference

Nomogram for working out diameter of conveying
pipe line for different capacities for air / material ratio = 50:1

DEOLALKAR CONSULTANTS

Nomo no.:	4-7-2 page 2
Title:	Nomogram for working out diameter of conveying pipe line for different capacities for air/material ratio 50:1
Useful for:	design of air lift systems
Inputs:	1 capacities of air lift in tph
	2 air to material ratio 50 : 1
Output:	diameter of conveying pipe line in mm
Scale:	A : 1.5 cm = 50 tph
	B : specially constructed scale
	P : point of reference
How to use:	**Example:**
	Draw a line from point of reference P thru given capacity (200) on line A and extend it to meet line B and read pipe line size in mm ~ (415) on it.
	note: commercially available size nearest to the calculated value will be selected
Source:	base data from CPAG Manual
	nomogram constructed

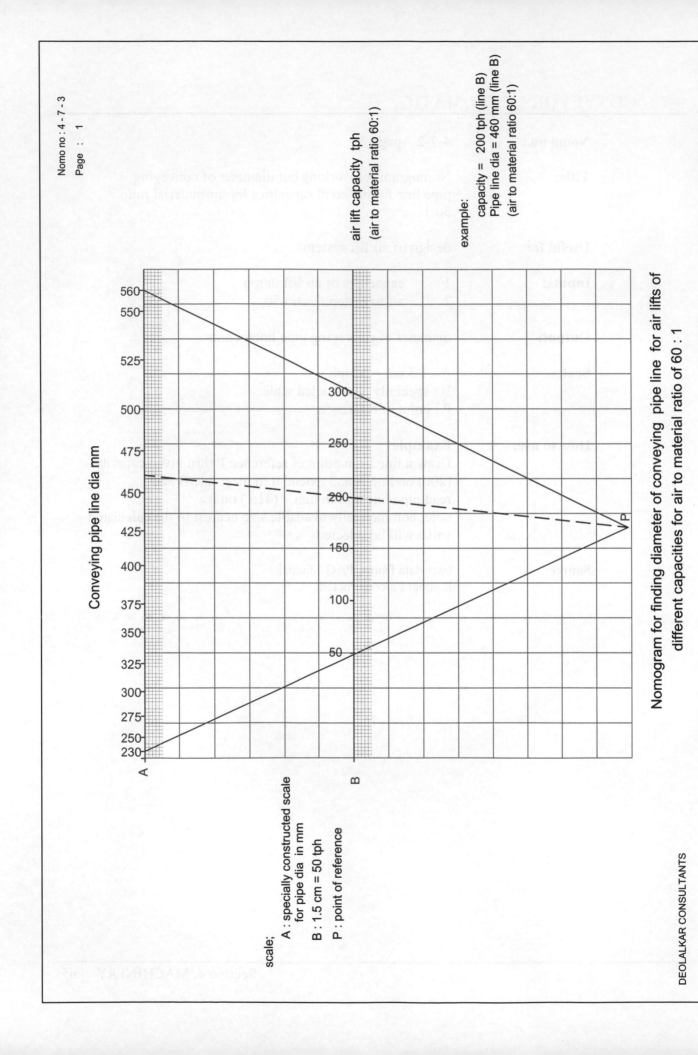

Nomogram for finding diameter of conveying pipe line for air lifts of
different capacities for air to material ratio of 60 : 1

air lift capacity tph
(air to material ratio 60:1)

example: capacity =   200 tph (line B)
           Pipe line dia = 460 mm (line B)
           (air to material ratio 60:1)

Conveying pipe line dia mm

scale;
A : specially constructed scale
    for pipe dia in mm
B : 1.5 cm = 50 tph
P : point of reference

DEOLALKAR CONSULTANTS

Nomo no.:	4-7-3 page 2
Title:	Nomogram for working out diameter of conveying pipe line for air lifts for different capacities for air to material ratio of 60:1
Useful for:	designing air lift systems
Inputs:	1 capacity of air lift in tph 2 air to material ratio 60 :1
Output:	diameter of conveying pipe line in mm
Scale:	A : specially constructed scale for dia. of pipeline B : 1.5 cm = 50 tph P : point of reference
How to use:	**Example:** Draw a line from reference point P thru the capacity in tph on line B (200) and extend to meet line A and read dia. of pipe line in mm (460) Note : commercially available size nearest to the size thus obtained would be used (Compare this with nomo 4-7-2) where pipe line size was 410 mm for air to material ratio of 50 :1)
Source:	base data from CPAG Manual nomogram constructed

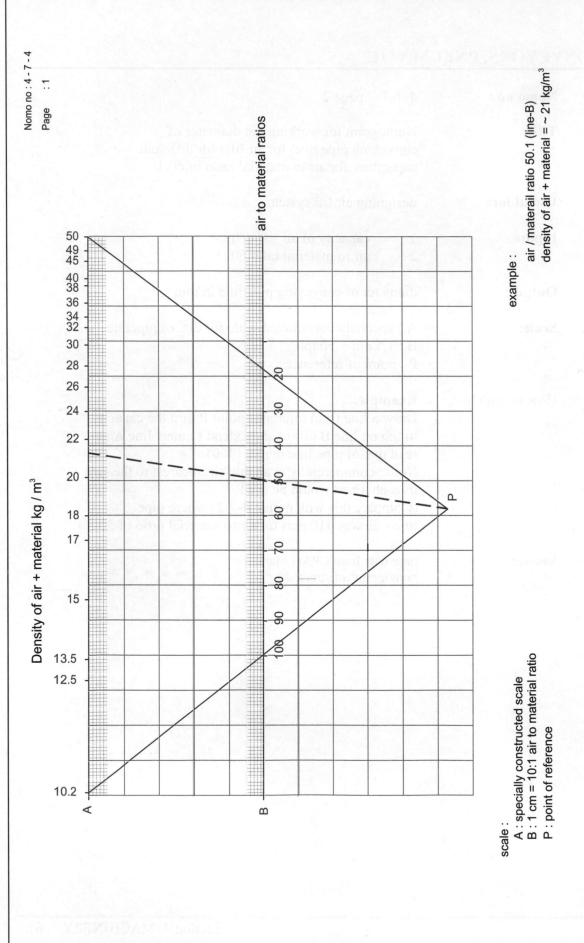

Density of air + material kg / m³

air to material ratios

example :

air / materail ratio 50.1 (line-B)
density of air + material = ~ 21 kg/m³

scale :
A : specially constructed scale
B : 1 cm = 10:1 air to material ratio
P : point of reference

(Note : this is valid of material densities between 0.7 to 1.2 t/m³)

Nomogram for working out densities of air + material mixtures for
different air to material ratios

Nomo no.:	4-7-4 page 2
Title:	Nomogram for working out densities of air + material mixtures for different air to material ratios
Useful for:	working out back pressure in air lift systems
Inputs:	1 air to material ratios 2 density of material conveyed (700-1200 kg/m^3) 3 density of air F.A.D. = 1.22 kg/ m^3
Output:	density of air + material for different air to material ratios Note: this nomogram is valid for bulk densities of 700-1200 kg / m^3
Scale:	A : specially constructed scale for density of air + material mixture B : 1 cm = 10 air to material ratio P : point of reference
How to use:	**Example:** air to material ratio 50 :1 join point P and 50 on line B and extend to line A read density, 21 kg / m^3
Source:	base data from CPAG Manual nomogram constructed

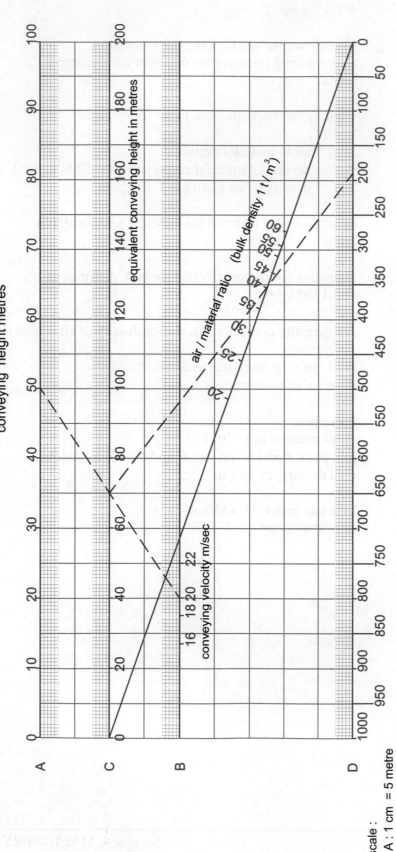

conveying height metres

A

C

equivalent conveying height in metres

B

air / material ratio (bulk density 1 t / m³)

60
55
50
45
40
35
30
25
20

16 18 20 22
conveying velocity m/sec

D

back pressure in mbar

example :

lift 50 m (line A), velocity, 20m / sec - (line B)
equivalent distance, 70 m (line C)
air material ratio = 40:1 (0-0)
back pressure = 190 mbar

Nomogram for finding back pressure in air lifts

scale :

A : 1 cm = 5 metre

B : specially constructed

C : 1 cm = 10 metres

D : 1 cm = 50 mbar back pressure

Nomo no.:	4-7-5 page 2
Title:	Nomogram for finding back pressure in air lifts
Useful for:	System design for Air Lifts

Inputs:

1	conveying height in m	0-100 m
2	conveying velocities	18-22 m/sec
3	air to material ratios	20-60
4	densities of material conveyed	0.7-1.2 t/ m^3

Output: Back pressure in mbar (without friction loss)

Scale:

A : 1 cm = 5 metres
B : specially constructed scale
C : 1 cm = 10 metres
D : 1 cm = 50 mbar back pressure

How to use:

Example:
1st step – work out equivalent conveying height
conveying height = 50 m (line A)
conveying velocity = 20 m/sec (line B)
equivalent conveying height = 70 m (line C)
2nd step - let bulk density of material be 1 t/m^3
let air to material ratio be 40:1
From 70 m on line B draw a line thru 40 on lin 0-0
to meet line D and read back pressure ~ 190 mbar.
This is valid for bulk densities ranging between
0.7-1.2 t/m^3

Source:

Base data from CPAG Manual
nomogram constructed

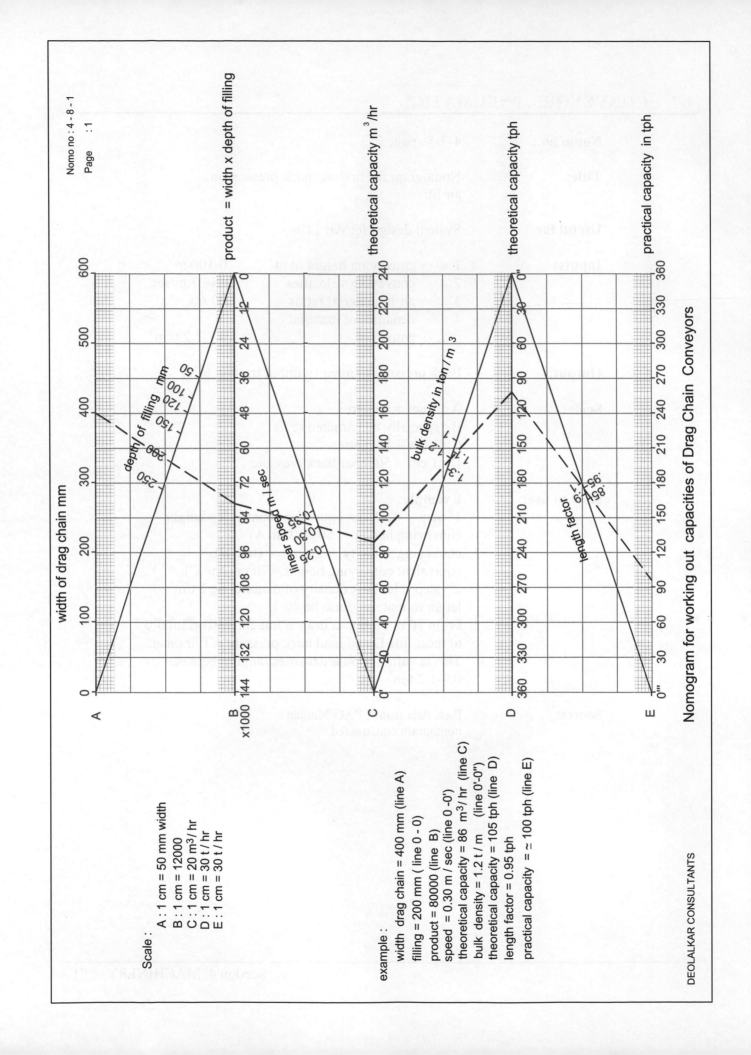

product = width x depth of filling

theoretical capacity m³/hr

theoretical capacity tph

practical capacity in tph

width of drag chain mm

depth of filling mm

linear speed m / sec

bulk density in ton / m³

length factor

Scale :

A : 1 cm = 50 mm width
B : 1 cm = 12000
C : 1 cm = 20 m³ / hr
D : 1 cm = 30 t / hr
E : 1 cm = 30 t / hr

example :
width drag chain = 400 mm (line A)
filling = 200 mm (line 0 - 0)
product = 80000 (line B)
speed = 0.30 m / sec (line 0 -0')
theoretical capacity = 86 m³/ hr (line C)
bulk density = 1.2 t / m   (line 0'-0")
theoretical capacity = 105 tph (line D)
length factor = 0.95 tph
practical capacity = ≃ 100 tph (line E)

Nomogram for working out capacities of Drag Chain Conveyors

DEOLALKAR CONSULTANTS

Nomo no.:	4-8-1 page 2
Title:	Nomogram for working out capacities of Drag Chain Conveyors
Useful for:	Sizing Drag Chain Conveyors used for conveying clinker in cement plants
Inputs:	1 width of drag chain mm - 100-600 mm 2 depth of filling mm - 80-250 mm 3 speed of drag chain m/sec - 0.25 – 0.35 m/sec 4 bulk density of clinker - 1.1-1.4 t/m^3 5 length of drag chain metres - 0-60 m
Output:	Capacity of drag chain in tph
Scale:	A : 1 cm = 50 mm width B : 1 cm = 12 × 1000 product – width × depth C : 1 cm = 20 m^3/hr capacity D : 1 cm = 30 tph capacity E : 1 cm = 30 tph capacity after length factor
How to use:	**Example:** drag chain width 400 mm (line A) depth of filling 200 mm on line 0-0 product = 80 × 1000 (line B) From this point join 0.3 on line 0-0′ to meet line C ; read capacity (~ 86 m^3/hr); Bulk density 1.2 t/m^3 on line 0′-0″ Theoretical Capacity in tph 105 tph (line D) Length factor 0.95 on line 0′-0‴ Practical capacity = ~ 100 tph (line E)
Source:	Base data from Ottolabahn nomogram constructed

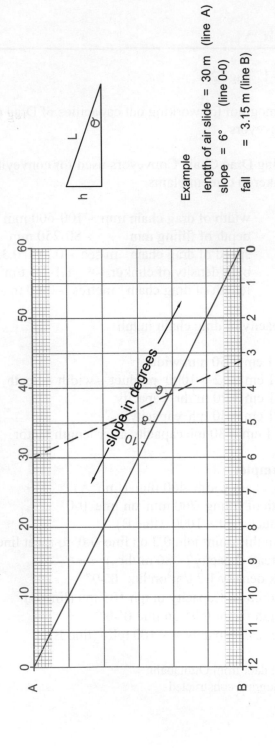

Example

length of air slide = 30 m   (line A)

slope = 6°   (line 0-0)

fall   = 3.15 m (line B)

Actual length -L - of air slide in metres

slope in degrees

fall 'h' in metres

Nomogram for working out fall in height of
air slides of different lengths & slopes

Scale : A :1 cm = 5 m

B : 1 cm = 1 metre

Nomo no.:	4-8-2 page 2
Title:	Nomogram for working out fall in height of air slides of different lengths and slopes
Useful for:	for working out head room required for installing airslides in layouts
Inputs:	1 Actual length of air slide in meters 2 slopes of air slides in degrees
Output:	fall in height in meters

$$h \quad \boxed{\overset{L}{}} \to \theta$$

Scale:	A : 1 cm = 5 metres B : 1 cm = 1 metre
How to use:	Draw a line from given length of air slide on line A thru designed slope on line 0-0 to meet line B and read fall in height in metres on it **Example:** length of air slide = 30 m (line A) slope = 6 ° (line 0-0) fall in height = ~ 3.15 m (line B)
Source:	constructed

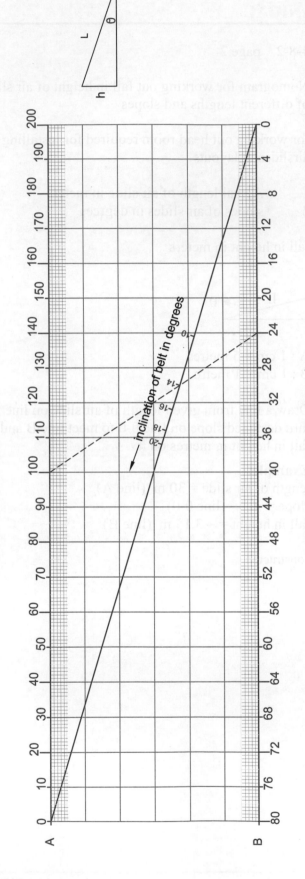

Length of belt conveyor in metres   L

Rise of conveyor in metres   h

inclination of belt in degrees

example : from 100 on line A draw a line 14° thru inclination on line 0-0
to extend to meet line B, read rise ~ 24.2 metres

Nomogram for working out rise of inclined belt conveyors for differents degrees of inclination.

Scale : A :1 cm = 10 metre
        B : 1 cm = 4 metre

DEOLALKAR CONSULTANTS

Nomo no.:	4-8-3 Page 2
Title:	Nomogram for working out rise of inclined belt conveyors for different degrees of inclination
Useful for:	designing layouts
Inputs:	1 length of conveyor in metres L 2 inclinations of conveyors in degrees θ
Output:	rise in and hence point of discharge of belt conveyor h

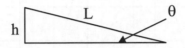

Scale:	A : 1 cm = 10 metres B : 1 cm = 4 metres
How to use:	draw a line from the given length of belt on line A thru angle of inclination on line 0-0 and extend it to meet line B and read rise of discharge point

Example:
length of belt conveyor = 100 m   (line A)
inclination   = 14 °   (line 0-0)
rise of belt   = ~ 24 m   (line B)

Source:	constructed

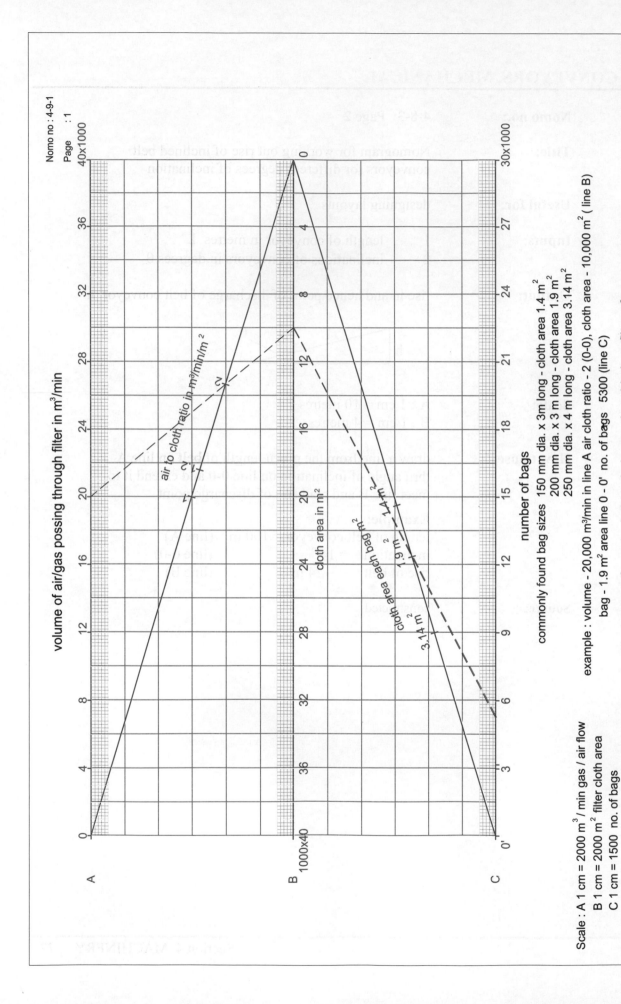

volume of air/gas passing through filter in m³/min

air to cloth ratio in m³/min/m²

cloth area in m²

cloth area each bag m²

3.14 m²
1.9 m²
1.4 m²

number of bags

commonly found bag sizes 150 mm dia. x 3m long - cloth area 1.4 m²
200 mm dia. x 3 m long - cloth area 1.9 m²
250 mm dia. x 4 m long - cloth area 3.14 m²

example : volume - 20,000 m³/min in line A air cloth ratio - 2 (0-0), cloth area - 10,000 m² (line B)
bag - 1.9 m² area line 0 - 0' no. of bags 5300 (line C)

**Nomogram for calculating cloth area in m² in bag filters for
different volumes and air / cloth ratios and to find number of bags**

Nomo no : 4-9-1
Page   : 1

Scale : A 1 cm = 2000 m³ / min gas / air flow
B 1 cm = 2000 m² filter cloth area
C 1 cm = 1500 no. of bags

DEOLALKAR CONSULTANTS

Nomogram no.:	4-9-1 page 2
Title:	Nomogram for calculating cloth area in m^2 in bag filters for different volumes and air to cloth ratios and to find number of bags
Useful for:	arriving at size of bag filter and number of bags
Inputs:	1 volume of air /gas passing thru bag filter 2 air to cloth ratios in m^3/min/ m^2 3 area /bag for bags of different sizes
Outputs:	1 filter cloth area 2 number of bags
Scale:	A : 1 cm = 2000 m^3/min gas flow B : 1 cm = 2000 m^2 filter cloth area C : 1 cm = 1500 no. of bags
How to use:	draw a line from pt. on line A, showing volume of gas to be treated thru recommended air-cloth ratio on line 0-0 and extend to meet line B. read net cloth area in m^2. From this point draw a line thru the cloth area per bag on line 0-0' to meet line C and read number of bags for net cloth area.

Example:

Volume of gas :	20000 m^3/min (line A)
Air to cloth ratio :	2 (line 0-0)
Filter cloth area :	10000 m^2 (line B)
Cloth area per bag :	1.9 m^2 (line 0-0')
Number of bags :	~ 5300 (line C)

Source:	constructed

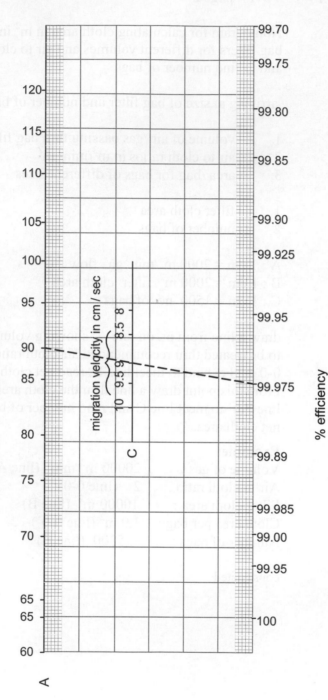

Scale : A: Specially constructed scale (SCA)
        B: Specially constucted for efficiency of ESP in %

example : n efficiency = 99.975 (line B)
          migration velocity ; 9 cm / sec (line C)
          SCA = 90 m^2/ m^3 /sec

Nomogram for calculating specific collecting area (SCA) for ESPs
for kiln exhaust gases for given efficiency & migration velocity

Nomo no.:	4-10-1 page 2
Title:	Nomogram for calculation of Specific Collection Area (SCA) for ESPs for kiln exhaust gases for given efficiency and migration velocity
Useful for:	sizing ESPs (Electro Static Precipitators)
Inputs:	1 efficiencies in % 2 migration velocities in cm / sec
Output:	SCA in $m^2/ m^3 /$ sec
Scale:	A : specially constructed scale for SCA in $m^2/ m^3 /$ sec B : specially constructed scale for efficiencies in % C : specially constructed scale for migration velocities in cms / sec Note : the scales have been constructed from curves furnished by Lurgi's for application of ESPs for kiln exhaust gases
How to use:	draw a line showing desired efficiency of esp on line B thru appropriate migration velocity on line C to meet line A and read SCA on it

Example:
desired efficiency = 99.975 % (line B)
recommended migration velocity = 9 cm / sec (line C)
required SCA = ~ 90 $m^2/ m^3 /$ sec   (line A)

Source:	Base data from Lurgi nomogram constructed

APPENDIX 1

In the nomograms included in the Book there are many in which log scale has been used. It is not practical to mark the scale in fractions However it is often required to read values that are not marked. The following table furnishes means to do so.

For example,

when scale used is 10 cms = 10-100, and it is required to read 14 on it, then 1.4 = 14.6 mm. Mark 14.6 mm from 10 to read 14

Table for plotting logarithmic scale

Logarithmic scale for 1-10, 10-100, 100-1000 etc.

number	log to base =10	4cms =	5 cms =	7 cms = mms from 0	10 cms =	20 cms =
1	**0**	**0**	**0**	**0**	**0**	**0**
1.1	0.041	1.7	2.1	2.9	4.1	8.3
1.2	0.079	3.2	4.0	5.5	7.9	15.8
1.3	0.114	4.6	5.7	8.0	11.4	22.8
1.4	0.146	5.8	7.3	10.2	**14.6**	29.2
1.5	0.176	7.0	8.8	12.3	17.6	35.2
1.6	0.204	8.2	10.2	14.3	20.4	40.8
1.7	0.230	9.2	11.5	16.1	23.0	46.1
1.8	0.255	10.2	12.8	17.9	25.5	51.1
1.9	0.279	11.2	13.9	19.5	27.9	55.8
2	**0.301**	**12.0**	**15.1**	**21.1**	**30.1**	**60.2**
2.1	0.322	12.9	16.1	22.6	32.2	64.4
2.2	0.342	13.7	17.1	24.0	34.2	68.5
2.3	0.362	14.5	18.1	25.3	36.2	72.3
2.4	0.380	15.2	19.0	26.6	38.0	76.0
2.5	0.398	15.9	19.9	27.9	39.8	79.6
2.6	0.415	16.6	20.7	29.0	41.5	83.0
2.7	0.431	17.3	21.6	30.2	43.1	86.3
2.8	0.447	17.9	22.4	31.3	44.7	89.4
2.9	0.462	18.5	23.1	32.4	46.2	92.5
3	**0.477**	**19.1**	**23.9**	**33.4**	**47.7**	**95.4**
3.1	0.491	19.7	24.6	34.4	49.1	98.3
3.2	0.505	20.2	25.3	35.4	50.5	101.0
3.3	0.519	20.7	25.9	36.3	51.9	103.7
3.4	0.531	21.3	26.6	37.2	53.1	106.3
3.5	0.544	21.8	27.2	38.1	54.4	108.8
3.6	0.556	22.3	27.8	38.9	55.6	111.3
3.7	0.568	22.7	28.4	39.8	56.8	113.6
3.8	0.580	23.2	29.0	40.6	58.0	116.0
3.9	0.591	23.6	29.6	41.4	59.1	118.2

Logarithmic scale for 1-10, 10-100, 100-1000 etc.

number	log to base =10	4cms =	5 cms =	70 cms =	10 cms =	20 cms =
				mms from 0		
4	**0.602**	**24.1**	**30.1**	**42.1**	**60.2**	**120.4**
4.1	0.613	24.5	30.6	42.9	61.3	122.6
4.2	0.623	24.9	31.2	43.6	62.3	124.6
4.3	0.633	25.3	31.7	44.3	63.3	126.7
4.4	0.643	25.7	32.2	45.0	64.3	128.7
4.5	0.653	26.1	32.7	45.7	65.3	130.6
4.6	0.663	26.5	33.1	46.4	66.3	132.6
4.7	0.672	26.9	33.6	47.0	67.2	134.4
4.8	0.681	27.2	34.1	47.7	68.1	136.2
4.9	0.690	27.6	34.5	48.3	69.0	138.0
5	**0.699**	**28.0**	**34.9**	**48.9**	**69.9**	**139.8**
5.1	0.708	28.3	35.4	49.5	70.8	141.5
5.2	0.716	28.6	35.8	50.1	71.6	143.2
5.3	0.724	29.0	36.2	50.7	72.4	144.9
5.4	0.732	29.3	36.6	51.3	73.2	146.5
5.5	0.740	29.6	37.0	51.8	74.0	148.1
5.6	0.748	29.9	37.4	52.4	74.8	149.6
5.7	0.756	30.2	37.8	52.9	75.6	151.2
5.8	0.763	30.5	38.2	53.4	76.3	152.7
5.9	0.771	30.8	38.5	54.0	77.1	154.2
6	**0.778**	**31.1**	**38.9**	**54.5**	**77.8**	**155.6**
6.1	0.785	31.4	39.3	55.0	78.5	157.1
6.2	0.792	31.7	39.6	55.5	79.2	158.5
6.3	0.799	32.0	40.0	56.0	79.9	159.9
6.4	0.806	32.2	40.3	56.4	80.6	161.2
6.5	0.813	32.5	40.6	56.9	81.3	162.6
6.6	0.820	32.8	41.0	57.4	82.0	163.9
6.7	0.826	33.0	41.3	57.8	82.6	165.2
6.8	0.833	33.3	41.6	58.3	83.3	166.5
6.9	0.839	33.6	41.9	58.7	83.9	167.8
7	**0.845**	**33.8**	**42.3**	**59.2**	**84.5**	**169.0**
7.1	0.851	34.1	42.6	59.6	85.1	170.3
7.2	0.857	34.3	42.9	60.0	85.7	171.5
7.3	0.863	34.5	43.2	60.4	86.3	172.7
7.4	0.869	34.8	43.5	60.8	86.9	173.8
7.5	0.875	35.0	43.8	61.3	87.5	175.0
7.6	0.881	35.2	44.0	61.7	88.1	176.2
7.7	0.886	35.5	44.3	62.1	88.6	177.3
7.8	0.892	35.7	44.6	62.4	89.2	178.4
7.9	0.898	35.9	44.9	62.8	89.8	179.5
8	**0.903**	**36.1**	**45.2**	**63.2**	**90.3**	**180.6**
8.1	0.908	36.3	45.4	63.6	90.8	181.7
8.2	0.914	36.6	45.7	64.0	91.4	182.8
8.3	0.919	36.8	46.0	64.3	91.9	183.8
8.4	0.924	37.0	46.2	64.7	92.4	184.9
8.5	0.929	37.2	46.5	65.1	92.9	185.9
8.6	0.934	37.4	46.7	65.4	93.4	186.9
8.7	0.940	37.6	47.0	65.8	94.0	187.9
8.8	0.944	37.8	47.2	66.1	94.4	188.9
8.9	0.949	38.0	47.5	66.5	94.9	189.9

logarithmic scale for 1-10, 10-100, 100-1000 etc.

number	log to base =10	4cms =	5 cms =	70 cms = mms from 0	10 cms =	20 cms =
9	0.954	38.2	47.7	66.8	95.4	190.8
9.1	0.959	38.4	48.0	67.1	95.9	191.8
9.2	0.964	38.6	48.2	67.5	96.4	192.8
9.3	0.968	38.7	48.4	67.8	96.8	193.7
9.4	0.973	38.9	48.7	68.1	97.3	194.6
9.5	0.978	39.1	48.9	68.4	97.8	195.5
9.6	0.982	39.3	49.1	68.8	98.2	196.5
9.7	0.987	39.5	49.3	69.1	98.7	197.4
9.8	0.991	39.6	49.6	69.4	99.1	198.2
9.9	0.996	39.8	49.8	69.7	99.6	199.1
10	1.000	40	50.0	70.0	100.0	200.0

APPENDIX 2

PROCEDURE FOR USING THE NOMOGRAMS

1   In the text attached to each nomogram procedure for using it has been given.

   By way of example one input and corresponding output for a given variable has been illustrated by a blue line.

2   Open a nomgram on screen. Read inputs, variables and outputs and their corresponding scales

3   size the nomogram so that both input and output scales are visible

4   type 'line' against command at bottom left of the autocad drawing. Press 'enter.

5   message select first point will appear in that place. Move the curser which takes the shape of a point to the

   Case 1

   desired point on input line and move it thru the desired point on the line showing variable and extend to meet output line.

   Press enter to complete the operation.

   Case 2

   Put curser on point of reference P and draw line thru value of input and extend to line showing output

6   enlarge the nomogram so that scales enlarge and the values can be read easily.

7   all scales have been divided into tenth divisions for example 1 cm is divided into mm divisions thus if 1 cm represents 100 units, each mm division reprsents 10 units.

8   Many nomograms have log scales. For example values 1-10 , 10-100, 100-1000 etc,. will be shown by say 5, 7, 10 or 20 cms in a nomogram Appendix 1 furnishes a table with the help of which, fractional values can be read.

   For example say, 10 cms represent log scale 1 –10,

   The log scale has been marked to read integer values from 1 to 10.

   The centimeter scale has been marked in mm divisions.

   Suppose it is required to read a value 1.2 on this scale.

   See table in Appendix 1

   Choose column marked 10 cms

   Choose 1.2 in the first column and read horizontally in column for 10 cms i.e. 7.9 mms.

   Measure ~ 8 mms from 0 (starting point 1) and the point will indicate value 1.2 on the given log scale.